MORE THAN *JUST A* DOG

SIMON WOOLER

MORE THAN JUST A DOG

Understanding, loving and living with dogs. An essential guide for humans

While the author of this work has made every effort to ensure that the information contained in this book is as accurate and up-to-date as possible at the time of publication, medical and pharmaceutical knowledge is constantly changing and the application of it to particular circumstances depends on many factors. Therefore it is recommended that readers always consult a qualified veterinary specialist for individual advice. This book should not be used as an alternative to seeking specialist veterinary advice which should be sought before any action is taken. The author and publishers cannot be held responsible for any errors and omissions that may be found in the text, or any actions that may be taken by a reader as a result of any reliance on the information contained in the text which is taken entirely at the reader's own risk.

HarperElement
An imprint of HarperCollins*Publishers*
1 London Bridge Street
London SE1 9GF

www.harpercollins.co.uk

HarperCollins*Publishers*
Macken House, 39/40 Mayor Street Upper
Dublin 1, D01 C9W8, Ireland

First published by HarperElement 2025

3

© Simon Wooler 2025

Simon Wooler asserts the moral right to
be identified as the author of this work

A catalogue record of this book is
available from the British Library

ISBN 978-0-00-870748-4

Printed and bound in the United States

All rights reserved. No part of this publication may be reproduced, stored in a retrieval system, or transmitted, in any form or by any means, electronic, mechanical, photocopying, recording or otherwise, without the prior written permission of the publishers.

This book is dedicated to the two most important people in my life.

The first is my Mum. She has been there for me with unabating support through the bad decisions I have made in my life, and immeasurable cheerleading through the good ones. As you will see in the coming pages, she has helped selflessly and often with bravery. Thank you, Mum.

The second is my partner, Nina Cooper. Nina turned one seemingly bad idea, starting a small dog business, into a good one. She kept us afloat through the lean times of starting a small business and has contributed more than I can count in time and consummate writing skill to everything Sociable Dog does. Without Nina this book would not have made it beyond an idea. She persuaded me that I had it in me and has been my 'editor-in-chief', looking kindly over my shoulder throughout.

Thank you to two utterly remarkable women.
I love you both.

CONTENTS

Introduction ... 1

CHAPTER 1: The Dog That Made Me ... 5
A little insight into how, why and when the adventure began. Spoiler: It was a dog's fault.

CHAPTER 2: The Forming of a Beautiful Friendship ... 17
A dive into how we as a species became long term buddies and associates.

CHAPTER 3: The Learning Curve ... 29
The real way to build a beautiful relationship. More spoilers: It's not about pack leadership or dominance.

CHAPTER 4: Nature and Nurture ... 53
The old conundrum, but the truth may be unexpected.

CHAPTER 5: What on Earth Are You Talking About? ... 61
If only we could talk to the animals ...

CHAPTER 6: Busting the Myths   87
From 'pack leadership' to 'training with treats is bribery': let's set the record straight.

CHAPTER 7: The Accidental Trainer   103
We're not all born to train: getting the most out of the everyday routine.

CHAPTER 8: The Power of Play   113
Even more reasons for dogs and humans to have fun.

CHAPTER 9: When Good Intentions Go Bad   125
What to do when our dogs get the better of us.

CHAPTER 10: The Fearful Dog   141
Helping the dogs who find the world difficult to navigate.

CHAPTER 11: The Reactive Dog   171
Understanding the motivators for the dogs that seem always angry.

CHAPTER 12: Great Expectations   199
Making your perfect canine match.

CHAPTER 13: The Rescuers   215
The matchmakers trying to give dogs another chance at love.

CHAPTER 14: Significant Others 239
Vets, behaviourists and trainers – who else is in a dog's life?

CHAPTER 15: Life Changes 255
How our dogs negotiate changes in their lives from new friends to loss and grief.

CHAPTER 16: In Conclusion 283
Thoughts on making and keeping the relationship special.

Endnotes 291
Acknowledgements 297
Index 301

INTRODUCTION

We've been hanging out with dogs for millennia. We share our homes, our time, our sofas and sometimes our beds, yet we still often look at each other across the species divide baffled and bemused, yet utterly devoted to the team.

Exactly how dogs and humans came together we can't be 100 per cent sure – no one was writing anything down or capturing the moment on their iPhone. But, as best as we can ascertain, this all started off as a loose joint venture for mutual survival. A kind of 'you scratch my back and I'll watch yours' arrangement. And very successful it was – where other species disappeared into extinction, we survived and thrived.

Of course, humans being humans, we couldn't just leave it there. We had to claim mastery and mess with their gene pool. Over the years, we have fixated on our superior qualifications to be the leaders of the pack and then anguished about whether we can actually stay in control. We worry that, given half a chance, that goofy Labrador will, apparently, establish dominance, claim the bed and the fridge and leave us with the kibble and the kennel.

We've put our dogs to work and sent them to war. We've dressed them in costumes and taught them to dance. And they've stuck with us, which is, frankly, sometimes more than we deserve. Yet after thousands of years of evolving together, we still manage to misunderstand and misinterpret their needs and motivations and confuse them with ours.

This book is an attempt to unpack what's at the heart of this very special relationship and help bridge this gap in understanding. It's based on a good few years geeking out on behavioural science and many more working with troubled dogs and the people who love them. It looks at our journey together, how we can interpret their communication and get involved in their play. Most importantly, it explains how dogs learn, which is at the very heart of living together in harmony.

This isn't a training 'how to' guide or a history book, though you'll find elements of both those in these pages. It's an *understanding* book for anyone who shares my love and obsession with dogs.

It's taken a few highways and byways since I first flexed my fingers on the keyboard (or at least the two that have taken most of the load) as I've explored the latest academic and citizen science. It's brought back memories of dogs I've worked with and loved and tells some of their stories. It has, without doubt, brought me some smiles along the way, which I hope you'll share too.

INTRODUCTION

I'm unapologetic about sweeping away some of the stale old myths and challenging misconceptions, but I am prepared to explain why. If this provides reassurance or parts the clouds of human-canine mutual incomprehension to offer a few 'aha' moments of enlightenment, then I'll consider it job done.

Most of all, this book celebrates everything we share in this extraordinary bond, which is both an anchor through life's changes and a source of endless, unexpected joy.

CHAPTER 1

THE DOG THAT MADE ME

His look was hard, cold, distant. Not aggressive – more cynical, resigned, expecting nothing good. If you've ever had anyone look at you that way, you know that they probably aren't best mate material. Beers down the pub and swapping family photos won't be on the cards anytime soon. In fact, probably not ever. You would need a really good reason to try and make any kind of connection, and even that wouldn't be easy.

The dog giving me that long look was a hefty 45kg of Rottweiler with a wonky jaw and a stump of a tail. He'd been named Tyson, which tended to set expectations. His gaze followed me through the mesh of the kennel at the shelter. It wasn't welcoming.

He'd had a couple of shots at adoption before but had been returned both times with a few more notes added to his bite history rap sheet. Only one of the handlers could approach him with confidence so his opportunities to get out and about were limited.

I've always been a bit of a sucker for the underdog or the overlooked. Maybe that's what kept me standing there. But then, there it was. Or was it? Just a hint of curiosity in his eye … maybe not. But still.

'Can I walk him?' I asked.

'Are you sure?'

Tyson's handler went into the kennel and emerged with this Goliath of a dog. I walked alongside them for a while and then asked to take over the lead. The handler was, shall we say, surprised at my enthusiasm to get stuck in and not without reason. Tyson was selective about his walking companions. I already knew that. He tolerated some human company, but absolutely no canine. And so it was with these warnings about other dogs and 'some' other people fresh in my mind that I took over, my hand sweating slightly as it maintained a vice grip on the lead. The lead itself was short and made of heavy chain links. A result of lessons learned, perhaps?

Curiously, the gentleness of the handover seemed to cause the big man to overlook the newness of his current escort. He seemed calm enough. But there were a few indicators of the way things could be. Those people he knew on the staff split and scattered when they saw him. That was a tip-off – it probably should have been a red flag. But then there was that soft spot I had for the misunderstood, waving 'hello' to the improbability and insanity of the situation.

The first time Big T wanted to tell another dog what he thought about meeting (basically, that he wasn't minded so to do) was something of an eye-opener. I had been warned, but I wasn't really prepared. I don't know how much pressure is brought to bear on a line when 45kg of Rottweiler comes to the end of it at full speed but I have a sense memory of it in my shoulder to this day. Note to self: do not hang around very long if other dogs are anywhere in the vicinity. It was also advisable not to get too close to unfamiliar people. Unfamiliar to Big T that is. He didn't give two poodles' tails if they were familiar to me. As far as he was concerned they were trouble, and no one was going to persuade him otherwise.

I probably have an over-romanticised memory of that first walk at the shelter, but in my head it went OK, apart from the need to keep him away from any people or dogs. In nearly every step you could feel Big T's tension, his wariness, his suspicion. He would occasionally relax into a sniff or glance back, in what I hoped was acknowledgement towards me, the woefully inadequate creature hanging on to the other end of the lead, willing to get to the end of this first outing together without disaster but at the same time resolving to be 'the one'. Perhaps we weren't quite bonding yet (of course we weren't) but there was the distant promise of a connection and that would do for now.

After that first walk, I spent three weeks visiting him. I sat in his kennel, reading while he inspected me and got used to my smell and my movements. Then, one day, he came and sat next to me and that was that. I was IN. Let the journey begin, whatever it would turn out to be. The shelter staff were more than accommodating about my decision. Or at least that was how I interpreted their mood at the time. In truth, my guess would be they were relieved and not a little bemused, having probably resigned themselves to having the 'Tyson' with them as a permanent resident – or making the call on euthanasia. Nobody wanted to use the 'E' word, but it was sometimes one of life's realities in shelters with limited resources. I knew that I wanted to teach him to trust me, even if it took a lifetime. After all, I was almost certainly his last chance. I brought him home, renamed him Thomson and we embarked on an eight-year voyage of discovery together. He changed my life completely.

EARLY MONTHS: LOSSES AND GAINS

There were, of course, practical adjustments to be made. When you collect a 45kg Rottweiler with 'issues' in a Peugeot soft-top and try to buckle him into a seat belt you soon realise that you're going to have to make some changes to your life, if only because he (and I) looked ridiculous. The car was going to have to go. But much more of a wrench was the motorcycle – that would need to go too.

I loved that bike. It was 1000cc of exquisite Japanese engineering that demanded respect. When I first walked into the showroom for the test ride I was sure I wouldn't like it. I was a Kawasaki kid. I just wanted to be sure that all those raving about the latest Yamaha were wrong. They were not. That bike and I were meant for each other. I was the customer the designers had imagined, as they carefully crafted this badass of a bike. It was meant to be. Goodbye Kawasaki. It's been great but, well, what can I say? 'I'm sorry'? There are worlds of back roads to explore and from now on, it's going to be me and the Yamaha. But with the arrival of Thomson, a choice had to be made. It was my statement of intent, my announcement to the world. 'This dog is staying!' Besides this job was going to need a serious car. A car with nearly as much attitude as the new passenger. A car you had to climb into. So, the soft-top and superbike were swapped for a second hand 4x4 that could cope with Big Thomson's scale.

More fundamental was the decision, some time later, to call time on a career of more than twenty years as a sound engineer so I could spend the time with him and do what I could to help him adjust to the world. My outlook and my life changed completely. I went from a working world where everything operated at pace to one where I had to take everything slowly, one teetering, patient step at a time. From a world where success is counted by the end product, audible and obvious, to one where I was constantly looking for barely perceptible micro steps of improvement that I could build on.

Let's be honest: when you adopt a dog like Thomson, every day is usually 'a work in progress' until, one day, they sign out. The most you can do is make every day better than the last – or at least try to. I think I did that for him;

I hope I did. He was the dog that made me. He is why I am where I am and why, more than a decade on, he remains my guide, my inspiration and my conscience. It wasn't always easy or comfortable, but I don't regret it for one second. Not a millisecond.

But we're getting ahead of ourselves here. Let's go back to the beginning. When I started on the adventure with Thomson, I knew very little about dog behaviour other than what I'd seen on TV or heard from the guy down the pub with two Jack Russells. My up close and personal experience of dogs as a child was limited. I loved them without a doubt, but the full extent of my knowledge was a dippy and adorable English Springer Spaniel called Raz – the family pet while I was growing up and a companion in mischief. Raz was a far cry from this damaged soul that was to share my life and newly acquired 4x4. Back then, the idea was that you had to be the pack leader. You were warned to be constantly on the alert for any attempt by your dog to knock you off the top spot and take control of the household. It's a myth that's still sticking around today, but we'll come to that later.

In the early weeks and months, reality checks about the implications of taking on Thomson came thick and fast. It wasn't anything like straightforward arranging care for him while I was away. He took to some people well enough, but caution was the buzzword. I was still working as a sound engineer, which regularly took me away from home and so, in an act of utter selflessness, my Mum moved from

Yorkshire to Wiltshire to save the day. She is five foot two, but faced with this giant, half-dog, half-mountain, completely unfazed. That's a Yorkshirewoman for you. But still, it was a sacrifice that will stay in my heart until the day I die.

We adjusted to new routines and new challenges. One of the earliest was that Thomson needed to go to the vet for some jabs. The vet, not unreasonably, said he had to wear a muzzle and, of course, Thomson hadn't been trained to wear one. He certainly wasn't going to let me put one on him without a better reason than a trip to the vet.

After trying to get his nose in a muzzle, bucket-like in size, for an hour or more, panic was setting in. The time for the appointment was looming and the safety margin I had allowed myself for getting there was rapidly disappearing. Finally, I decided that the only thing I could do was head to the shelter where I was sure help would be on hand. And indeed it was. It took four of us: two to hold poor Thomson, one to wrap some bandage around his muzzle and another to deftly slip on the muzzle as the bandage was released and removed. Removing the muzzle after a, shall we say, dynamic visit, was another story altogether. Not as challenging as the application but nerve-shredding nonetheless.

You might now be getting an image of me: up for a challenge, perhaps; reckless, possibly; almost certainly single-minded and stubborn. Those traits, whatever your view of them, kept Thomson with me and not heading back to the shelter. I can't tell you how hard that first month was. On one occasion while I was away and my mum was 'in charge' of Thomson, I got a call explaining that he wouldn't let her down the stairs. This hadn't happened before and came as a surprise to both of us. It resolved eventually and, as with all the set-backs, didn't shake my mum's commitment to help him. Single-mindedness clearly runs in the genes.

On another occasion, a visitor to the house had ignored the large sign on the gate not to enter if 'the dog' was in the yard. Fortunately the gate wasn't easy to open, which gave me time to shout 'No' in what felt like slow motion, and grab Thomson's collar to pull him clear of the danger zone. Thomson, in appreciation, turned around and sunk his teeth into my wrist. Letting go immediately, he stood alongside me, eyeing the visitor suspiciously with a low, guttural growl. I asked what the visitor wanted. 'I've come to talk to you about household security,' he said. As blood slowly trickled down my hand and on to the paving slabs, I looked at him. 'I think I've got that covered.'

You'll be getting the picture now. But there was one last member of the family I haven't introduced. Thomson had been evaluated as 'safe with cats' while he was in the shelter, which was surprising since he was safe with very little else. That evaluation may well have been true. Indeed, he showed every sign of wanting to be chums with Oscar, my long-time and much, much loved ginger Tom. What those evaluations don't and can't tell you, however, is how your cat is going to feel about the new canine 'companion'. And as far as Oscar was concerned, the idea stank. Eventually, the feeling became mutual, which I suppose is all you can expect when the response to tentative, friendly overtures is repeated hissing and a swipe on the nose. In the first couple of months, I was frequently separating the two. Thomson would always come off worse. He would sit disconsolate with some newly acquired scratch across his nose or lip. There was only one master in this house.

THE KNOWLEDGE

There clearly needed to be some knowledge acquisition applied to this relationship or we were going to hit the skids. That search for information would take me through the uncharted landscape of dog training and behaviour. I may as well not beat about the bush: the industry is unregulated. Sure, there are plenty of well-educated practitioners but it's also possible to set up as a dog trainer (and even label yourself a behaviourist in the UK), take complex cases, charge money for your services and not have a Scooby Doo of what you are doing. The same is true of the available education and information on dog behaviour and training: the internet is awash with an unholy brew of well-meaning people with limited knowledge, well-qualified individuals with a drive to educate and less than scrupulous opportunists who see a money-making opportunity and little else.

I waded my way through much of it. I subjected Thomson to some of the pack leader 'bad' before getting to the good. It was only when a friend suggested I read Jean Donaldson's book *The Culture Clash* that I found what I had been looking for. A principle guiding how to train, founded firmly in scientific research. An evidence-based approach that quantifiably worked and respected the dog as an individual entitled to be treated respectfully and ethically.

Jean had, up until then, been running a six-week residential course for dog trainers at the San Francisco SPCA. It wasn't cheap and by the time I decided I was going to pull the trigger on enrolling, the course had ended. I was devastated. Nothing else I had seen came close.

Fortunately, Jean is an approachable person who engaged with me in an exchange of emails. She was developing an online version of the SFSPCA course that would be available worldwide and she would let me know when the launch would be.

I hounded Jean relentlessly. I was, I think, the first paying enrollee on the new course. From day one I saw it not as an expense, but an investment in what is widely seen as 'the Harvard of Dog Training'.

It is a tough course – two years of video coaching, webinars, tutorials, frequent assessments and a frankly mouth-drying, terrifying six-hour examination. I supplemented it with volunteering to get practical experience and we were expected to read extensively. But those two years gave me the confidence that I could accurately read a dog's body language, understanding what behaviour was motivated by fear and what was excitement or frustration. I could get owners to articulate in detail what they saw and experienced so I could draw up a plan. The Academy for Dog Trainers gave me both an understanding of behaviour and the mechanics to train efficiently and skilfully. It was the start of what would be the driving force of my life: a change of career and a thirst for learning that matched my huge sense of responsibility.

What had started out as the need to help one dog became a determination to help others.

And what of Thomson, the un-homeable Rottweiler, the dog that made me? Well, he taught me something every day of his eight years with me. And that wasn't just that you can't put a Rottie in a soft-top. In return I found the tools in the Academy and my ever-ongoing education to do right by him. He was never a party dog but he made *some* friends – friends of mine who gave their time and held their nerve to walk with him, some bringing their own dogs along. Step by carefully measured step, Thomson learned to like at least a select group of people and dogs, the most significant being my partner, Nina, and her little dog Woody. Nina was a stalwart from the day I met her. She did everything, and only what I asked her to do with Thomson. She would sit quietly with a cup of tea as Thomson viewed her from his favourite chair. She would later gently throw a ball for him to retrieve. A ball was the way to Big T's heart and Nina certainly found a place in it. As time went on, we started to introduce him to Woody. The inimitable Woody was 15kg of whirling, joyful Cocker Spaniel who believed, it seemed, that no one could resist him. He was largely right – though Thomson took a little longer than most. Eventually, however, Thomson came to learn that where Woody was, the love of his life, Nina, wasn't far behind. So, through the power of positive association, Woody worked his way into the big man's heart and gently softened it some more.

My search for the right way to help one fragile dog taught me about the real joy there can be when we start from the right place and commit to making a connection. Any relationship takes time, but we can make our lives with our dogs happier, safer and more enriched if we understand how they learn and put in the hours, the days and the months to teach, to play and to discover the world together.

CHAPTER 2

THE FORMING OF A BEAUTIFUL FRIENDSHIP

Our relationship with dogs has come a long way. For many of us, they're front and centre in almost everything we do. We've bent and moulded our lifestyles around them, organising our leisure so that we can bring them along for the ride. We take them to restaurants (choose from the dog menu) and pubs (dog beer in bottles, but not yet on tap). They come on holiday with us, whether it's a caravan, a cottage or a luxury hotel (dog bathtubs, activity programmes, massage and spas sometimes available). There are even dog-friendly cinema screenings where you can take your companion along to enjoy the latest blockbuster, though that feels like it has all the potential to be rather too high a decibel event for full enjoyment.

We want them at work, too. A Kennel Club survey found that half of us would be more likely to take a job that allowed dogs in the workplace[1] and HR magazine *People Management* reported that nearly three-quarters of employees would be happy to go along with that.[2]

Whether the dogs themselves might prefer to hang out at home, snuffling out a carelessly discarded shoe to chew, reorganising the cushions on the sofa and then having a romp in the park with a dog walker hasn't been reported. But, while, for some, being a workaday dog would be challenging, many would be very much up for the job, performance reviews and all. The fact is, many of us want our dogs with us wherever we are. And our dogs? Well, they're just fine with that.

When we can't have them at our side, we've got that covered, too: dog walkers, of course, day care, dog hotels (with TVs for added comfort) and in-home dog cams where we can check in, have a chat and offer the remote comfort of a treat.

They're becoming a central part of our human relationships: less 'Love me, love my dog' and more 'Do you love my dog more than me?' Celebrity spats over canine custody are splashed all over the media. Solicitors offer pet prenups and UK charity Blue Cross has a downloadable agreement, drafted with legal experts, to set out arrangements in the event of relationship breakdown.

For those who have resisted the appeal, the accommodations we make to our lives must seem absurd. 'It's just a dog, right?' And, in fairness, it's easy for the dog-obsessed among us to get this all overblown. A lot of the eyebrow-raising poll statistics are just PR grist for pet businesses and tabloid slow news days. But there is a profound truth at the heart: this is a special relationship.

Anyone who loves or has loved a dog will know that their simple presence heightens the senses on a familiar walk, makes a lazy sofa day feel warmer and more secure and any weekend, however grey and unpromising, the prompt for adventure. We are more likely to turn to our dogs in stressful situations than our parents or children – only romantic partners offer more comfort.[3] Dogs are our anchor in an uncertain world, as we are for them. Their companionship makes us more playful and joyful. Their loss flays our soul.

Like all good relationships, this one has been a long time in the making. Dogs have been nuzzling their way under our elbows and into our lives for tens of thousands of years so it's not surprising they've got it pretty much down to a fine art. But when did it all start and how did we get here?

WHEN HUMAN MET DOG

From the time, more than 20,000 years ago, when our ancestors were huddled round their campfire after a tough day hunter-gathering, our dogs' forebears started to move in. At that point, it seems unlikely that they ambled up with a wag and installed themselves comfortably in front of the fire, but they were almost certainly at the outer fringes of those early communities, watching the glow.

They were the wolves less wild – though only insofar as they had less fear of being in proximity to humans and were not as inclined to be aggressive or competitive over food (otherwise they would have been driven out). We can't be sure, of course, because neither species was big on writing things down at the time, but it's likely that they were savvy and opportunistic, spotting the potential for an easy meal of humans' leftovers. They thrived, bred and passed down their traits of tolerance to humans through the generations.

As a species, wolves appear ripe for this kind of relationship development, relying as they do on cooperative living within their own groups. They work collaboratively to hunt effectively and all play their part in feeding the young, even though only one pair will be the parents. Despite the mythology that has sprung up around alphas dominating their pack, wolves are, according to current scientific understanding, more communitarian than autocratic. We've just given them bad press.

But in understanding how those wolf-dogs on the fringes of the human campsite transformed into the dogs we share our lives with now is by no means simple.

The modern dog is very distinct behaviourally from the modern wolf, as well as physically. There is a school of thought that dogs are descended from an unknown and extinct wolf rather than the grey wolf we know today. (And there is other DNA in the mix; coyotes, jackals and foxes are all in the family tree, too.)

Decoding the dog genome, scientists found that though dogs share 99.9 per cent of mitochondrial DNA with

THE FORMING OF A BEAUTIFUL FRIENDSHIP

wolves, there were some important genetic differences that give pointers to how the relationship developed. In one chromosome, researchers found dogs had three genes linked to hypersociability.[4]

Whatever their precise antecedents, dogs have become particularly attuned to making attachments to humans and have proved skilled at learning the rules of cooperative living, unlike their lupine distant cousins. Researchers at the Wolf Science Center in Austria raised and trained four litters of puppies and four litters of wolf cubs. The wolves and the dogs were trained to follow basic commands and walk on leads. But the differences between modern dogs and wolves remained. If the researchers left a piece of meat on the table, a dog could be trained to 'leave it'. The wolf, even after seven years on the training programme, would look the human straight in the eye and take the meat. There were limits to their inter-species cooperation when it came to something really tasty.

Trying to understand the process of transformation from feral to friend continues to baffle and challenge scientists. It was at the heart of a long-running experiment structured by geneticist Dmitry Belyayev with silver foxes in Siberia, which began in 1959 and continues today. The foxes were selectively bred for tameness: being willing to be handled and keen to have contact with humans. The number of tame and tolerant foxes in each litter increased through successive generations and some familiar behaviours emerged. The scientists noted 'tail-wagging' in the fourth generation of pups.

Their appearance began to change, too. Pups appeared with floppy ears, rounder muzzles, shorter and sometimes curly tails, and different coloured coats. They lost the characteristic fox smell. Their levels of stress hormone decreased and serotonin increased. After forty years of selective breeding, the Siberian scientists had friendly foxes who enjoyed a belly scratch and who they could house-train and teach to sit on command. (They were never massively keen on leashes, but we can forgive them for that.)

Some of the conclusions of Belyayev's silver fox study have been questioned and challenged in recent years,[5] in the way that scientists do. Academics pointed out that the original foxes were not entirely wild. They had come from fur farms on Prince Edward Island in Canada so had experienced some human contact. Nevertheless, even those who question elements of Belyayev's hypothesis around the impact of purposeful domestication acknowledge the extraordinary value of the study. It gives us a window into how the wild – whether as a result of human intervention or 'self-domestication' through advantageous proximity to humans – became willing to be our BFFs.

THE MUTUAL TRADE AGREEMENT

The nature of the relationship with those early wolf-dogs seems to have been less one of master–subject and more a mutual trade agreement. Being close to humans provided access to food scraps. In return, the dog probably had a value as guard, alerting to potential intrusion and disrupting predators by their presence and movement, and as an ally in sniffing out potential game. These early wolf-dogs could have continued to live independently but chose to hang out with humans because it worked for them. We valued their presence and collaboration because they made us more successful, too. We could hunt more effectively and sleep more safely. This wasn't subjugation; this was the beginnings of an evolutionary profitable alliance.

There are indications, too, that even way back in those murky, beyond-written-history days, early humans had begun to form a real bond with their dogs. Burial sites in Sweden and Germany suggest that dogs were revered in death for their skill as hunters or treasured as family members.

Around the beginning of the fourteenth century, we started to inject more formality into the transformation process and started consciously breeding for specific skills, initially as hunting partners. But, even then, an element of that trusted trading relationship survived. The dog came along to facilitate the hunt, to track and tire the prey but, at the end of the day, was prepared to give up the prize for a share of the reward.

To this day we've continued to find useful jobs that dogs can help us out with and since the 1800s we've also bred for looks as well as function, so our dogs act as a status symbol or statement of identity simply by trotting alongside.

Over millennia, dogs and humans have evolved together ('convergent evolution' in the scientific jargon), which has strengthened our capacity to get along. That doesn't mean there isn't plenty of miscomprehension and bafflement to contend with in our day-to-day lives – we're still very different species and evolution doesn't account for everything in the individual relationships we have with the dogs we love. But over those many, many generations, dogs have made themselves both indispensable and agreeable to us. They've proved to be extraordinarily socially adept, skilled at modifying their behaviour to the 'rules' of the humans they live alongside to earn benefit for themselves.[6] So if you feel your dog is performing poorly at adapting to your rules, there's a simple lesson to be learned. You need to make the benefit clearer.

THE EMOTIONAL CONNECTION

Of course, this isn't the whole story. Our relationship with dogs is something much deeper than can be explained by a mutually beneficial trade of skills and resources. We think of them as our friends.

There's something similar in the approach of both dogs and humans to sociability with those outside their species. Dogs form an attachment to their caregivers; we provide emotional fulfilment to them and experience emotional benefits in return.[7] If we're familiar to them, dogs will look in our eyes (though long stares are still uncomfortable for both species). They inspect our faces and attempt to decode our expressions,[8] something they only do with humans – they don't look at other dogs in the same way.

There's some evidence that humans can understand the meaning behind a dog's different array of barks – even non dog owners can discriminate between a 'stranger alert' bark and a play bark when listening to a recording. And our dogs can take meaning from our tone of voice, even if vocabulary and syntax is a step too far. This is why, if an owner returns home to find their dog has chewed up a favourite trainer or removed the filling from the sofa cushions, they often think the dog looks 'guilty'. In fact, the dog is reading the displeasure in our expression or tone of voice and coming to the conclusion that, what had started as such a fun day, was about to go downhill.

MORE THAN JUST A DOG

Dogs mirror our yawns in the same way that humans do. Almost uniquely among animals, they will follow our gestures – looking in the right direction if we point at something. And when we play with them, they'll match our play movements, too.

They want to be close to us in unfamiliar environments or worrying, stressful situations (so, yes, it is absolutely OK to comfort your dog if they are frightened by something); the same reassurance isn't offered by the presence of another familiar dog.[9] Our presence also gives them more confidence in investigating the new.

We share a desire and a drive for physical contact, which creates a chemical reaction. Stroking our dogs increases the levels of the hormone oxytocin in both them and us, which is an important part of the bond between infants and mothers and in romantic relationships and enhances wellbeing. There's a chemical feelgood reaction to connecting with our dogs – or even just thinking about them. They can reduce the levels of the stress hormone cortisol, too.

But irrespective of what's going on with the hormones, being with dogs just does us good. Unlike other animals (including most humans), dogs stay playful well into adulthood and that encourages us and gives us permission to be playful, too. They are the essential prompt to get out there and take some exercise with all the associated health benefits. They're the antidote to social isolation, making even the most introvert of us more agreeable and sociable. Which dog owner hasn't gathered a circle of people on their

THE FORMING OF A BEAUTIFUL FRIENDSHIP

regular dog walking route that they're happy to engage in inconsequential chat, but probably know only by the dog's name?

The relationship between humans and dogs is a complex and powerful one that has been tens of thousands of years in the making and based firmly on the benefits we bring to each other. So let's stop worrying about trying to 'show who's boss' and tying ourselves up in knots about pack leadership and instead nurture and focus on all those positives. This is a partnership like no other.

CHAPTER 3

THE LEARNING CURVE: HOW YOUR DOG NAVIGATES THE WORLD

If you asked me to tell you what your dog is thinking, then the answer would be: I have no idea. That's sorted that out. Shall we all go down the pub?

We all like to believe we know what's going on in our dog's head when we see a particular expression on their face. We're inclined to construct whole narratives around challenging or confusing behaviour. It helps our own processing of the situation and makes it easier to empathise. Our dogs are family members so it's hardly surprising that we want to anthropomorphise and, quite literally, give our dog a voice.

The difficulties start when we fill the space with a hypothesis that turns out to be wildly off the mark. As rational as we try to be, the story we create for ourselves about why our dogs are acting in a certain way is often a long way from reality. If it's a problem behaviour and it gets worse, it can lead to confusion or frustration. We can find ourselves locked in a battle of wills because we've ascribed a motivation to our dog that exists, for the most part, in our imagination.

WHAT'S MY MOTIVATION?

While it's tempting to try and understand the *thinking*, too often that starts with assumptions of guilt or resentment, cunning or manipulation. None of which, in truth, is likely to be what's going on in the dog's head. Of course, I can't absolutely disprove that. But I do want to try to get you thinking in a different, more profitable way about your dog, a way that will bring you both even closer together. If we stop trying to guess what our dogs are thinking but look instead at their behaviour and what that tells us about their motivation, the better we'll understand each other. It also means we're better equipped to handle the challenging moments with equanimity – whether that is a dog barking at friends when they visit or snarking at the other neighbourhood dogs.

Dogs do what works: what works to keep them safe, their stuff safe, to gain nice stuff to eat and so on. Dogs essentially have a few simple goals in life: make good things happen and keep them happening; make bad things stop and ensure they don't happen again. How do I know this? The only way I can. By observing and evaluating behaviour. Does it increase or decrease and what has driven that change?

Consequences or outcomes drive behaviour change. It's that simple. It may not always be that simple to apply, but to get to an understanding of our dogs, it comes down to two questions:

Question: How do we know if our dogs are happy or not?
Answer: Through the way they behave. (NB Doing nothing is a behaviour.)

Question: How do we get our dogs to do something we want them to do?
Answer: By motivating them to do more of what we want and less of what we don't want by controlling access to rewards of reinforcers.

UNDERSTANDING THE DOG'S ECONOMICS

> 'It's the economy, stupid'
> – James Carville, advisor to Bill Clinton,
> 1992 presidential campaign

When you boil it down, dog training and behaviour is a matter of economics. There's a basic profit and loss assessment going on all the time. You, as the trainer, are evaluating what you need to spend to get the most bang for your buck (in behavioural terms) from your subject at the other end of the leash. Your dog, meanwhile, is trying to establish what he needs to do in order to access your 'cash': that stash of valuable treats hidden somewhere about your person or that game with a ball or a toy. His calculation is a simple one: what gets him paid and what doesn't get him paid. It's a volume game for him and each time he (at first) guesses wrong, is a profit opportunity lost. The spreadsheet of life, people. The spreadsheet of life.

Before we move on to the nuance of that last claim, let's just address the question of control and whether, in fact, your dog is a willing participant in this process or not.

To answer that, we should consider who has control here. Let's put aside, for the moment, the problem of changing emotional responses with fearful dogs and focus just on getting a change in behaviour from a dog who is not upset but motivated purely to gain access to some reinforcer or another. This kind of training is called operant conditioning. The dog *operates* on their environment in order to bring about an outcome or consequence. Their main objective is to influence events to gain something desirable or avoid something undesirable. They are the operator in question. It's *their* operation on the environment that drives the outcome. We, too, can operate the environment by manipulating the availability of any given reward or reinforcer. Your dog trades by offering a behaviour in exchange for the desirable reinforcer. If you are buying what he is selling, then it's a win-win. Both parties are happy.

So, do we have a coercive relationship here? Is this even a relationship between master and subordinate? I would argue not. What you have with your dogs is much more closely aligned to a partnership, a contractual agreement for mutual benefit. A contract that bridges the gap of species, solves the problem of mismatched languages and culminates in a perfectly balanced trading agreement between the best of friends. They love you and you them but, just like with teenagers, sometimes it's worth cutting a deal to keep the cogs of the family machine turning happily.

While getting inside the little black box of how our dogs think may be (for fans of *The Hitchhiker's Guide to the Galaxy*) a Babel fish too far,* while we don't have that fictional universal translator, we can make a decent assessment of how they are *feeling* at any given time.

You can tell when your dog is sad or frightened, feeling mischievous or joyful. I can tell that about our dog, too. There's little doubt about the love Ripley has for us or the delight she feels when she's on the beach chasing a ball. There's also no question that car journeys worry her. We can become very well attuned to our dogs' emotional states and they to ours. Because they are social animals we are as important to them as they are to us. If we know them well, we can tell how they feel.

But if we spend too much time trying to fathom what our dogs are *thinking*, we're missing the opportunity to improve their lives with us by understanding *how* they learn and so come to a better understanding of what is really behind what they do. Spoiler alert: it has nothing, absolutely nothing, to do with who is the pack leader.

---

* The Babel fish is a small yellow fish in Douglas Adams' *The Hitchhiker's Guide to the Galaxy* that, when inserted into the ear, translates from one spoken language to another.

LEARNING BY ASSOCIATION
(OR, IF THIS HAPPENS, THEN THAT TENDS
TO HAPPEN NEXT)

Ivan Petrovich Pavlov was a Russian scientist in the late nineteenth century who ran The Institute of Experimental Medicine in St Petersburg. Pavlov's thing was digestion and the subjects for his experiments (unfortunately rarely savoury) were dogs. Initially, his interest was in exploring the digestive workings of mammals, but he noticed something in the behaviour of the dogs in his lab that caught his attention. Their salivary reflex to food was time-travelling. At first, the food had to be in front of them, or at least they had to be able to smell it before they started to salivate. As time went by, however, the dogs started to respond to the sights and sounds of the food preparation. Then it was the sight of the lab assistants who brought the food. Later still, dogs were deliberately conditioned to salivate at the sound of a bell (or a metronome), which consistently predicted the delivery of food. It's that consistency that is the key to unlocking the alchemy of changing emotional responses in just about any animal on earth.

Pavlov's focus was on reflexes – how to facilitate an involuntary, physiological response to a given stimulus. Modern behaviour science has harnessed that essential discovery and used it to make the lives of people and other animals better by making associations. We have long known that you can help with fears and phobias by making careful and incremental associations between the fear-evoking stimulus and something that brings pleasure and positive feelings. This is known variously as classical, Pavlovian or respondent conditioning.

The majority of well-socialised family dogs will go through life taking the world in their stride. They'll meet people with pleasure and the neighbourhood dogs with a gentle sniff and perhaps an invitation to play. They'll eat their meals without fuss. Some may gulp it down like it was their last meal, while others graze delicately and politely, despite the proximity of the family cat eyeing the spoils of war. They might share the sofa gleefully with you, kicking their paws in the air until comfortable, allowing a sliver for you to perch on to watch the TV. They'll jump off the bed when asked so 'their' sheets can be changed and willingly relinquish even the rarest of finds when out on a walk if you offer to exchange it for a welcome, though less valuable, treat.

But this isn't true for all. Despite our best efforts and intentions, some dogs look at the world more with dread than excitement. They are unsure of the unfamiliar, be that people or dogs. They might be afraid of sudden changes to the 'norm' or see a stray plastic bag in the street as a danger. Their food or toys, their sleeping places or their people appear to them to be under constant threat from anyone or anything that approaches them. It is these dogs that I have spent much of the last fifteen years helping to find their way in the world. And for the record, that's just as much about being sensitive and sympathetic to the needs and worries of the humans who have made the call to get help for their dogs. Whichever way you slice it, this is about a family member in crisis and that deserves respect and kindness.

KEEPING IT SIMPLE

My teacher and mentor, Jean Donaldson, opens her remarkable course on dog behaviour with a simple question; a question you need to answer before you can possibly assume to start modifying any behaviour. That question is: 'Is my dog upset or not?'

It seems almost too simple, doesn't it? Of course you can tell if your dog is upset or not, right? But can you? Can you really? After all, you can't put your Pug on the couch and have a Freudian chat with him, can you? Moreover, surely you need to know more about what's going on before you can start formulating what must be a complex plan. But do you? And does it have to be?

A lot of questions follow on from that apparently simple one but don't worry, I'm not going to leave them unanswered.

Let us deal first with whether 'Is my dog upset or not?' is an adequate question. Naturally, I would argue it is because I was trained to think in those terms. But there's also a very good case for it and the following pages will lay out the case. (By the way, it's good news. We all have enough complexity in our lives without dealing with more when it comes to our dogs.)

Think back to earlier in this chapter when we talked about how dogs learn, the fundamental principles that govern the laws of behaviour. First there was operant conditioning, where dogs learn that they can influence outcomes using behaviour. They can effectively 'trade' behaviour to gain things they want or avoid things they don't want. Then there was Pavlovian or classical conditioning, where we can influence and modify emotional responses to all kinds of stimuli by making reliable associations.

Now think about the question again: Is my dog upset? If the answer is 'yes', you have two possible types of learning you could use to help her be less upset: Pavlovian or operant conditioning. Which would you choose? If you answered either, then you could well be right. But in a world where some right answers are more right than others, if you answered 'Pavlovian conditioning', you would be absolutely on the money. Why? Because we're trying to change emotions. We're not just trying to stop the poor dog being upset when faced with whatever the nemesis might be, but we're actively trying to make the dog pleased to see it. Our primary objective is to build positive associations between the problem and something that already positively rocks their world.

Now, you can train behaviours using operant conditioning and still get a Pavlovian effect to a stimulus, but it is a side effect. A side effect that you can exploit, but a side effect all the same. This can come in pretty useful. It's useful if the answer to our core question (Is my dog upset?) is ambiguous but it has other advantages too. Training your dog to perform an alternative behaviour to the one they are currently displaying (such as lunging, barking and snarking) does three things for us. The first is that it gives us control. If your dog is doing a nice close heel for you, he can't be lunging at the neighbour's Newfoundland. Secondly, it gives you a measure of how comfortable your dog is and whether, if upset, they are still under their fear threshold. Most upset dogs don't tend to perform trained behaviours, and they tend not to take food. Or if they do, they grab it along with your fingers and then quickly focus back on the problem. The third thing it does for you is that Pavlovian side effect I talked about above. 'If I see another dog I'm afraid of (problem), I get to do my thing (trained behaviour), which gets me paid (food). I like other dogs (Pavlovian side effect).' The whole point is the side effect. Cool! Right?

So if your dog is upset then you have some more tools in your arsenal to deal with. Both Pavlov and B. F. Skinner, the American psychologist widely viewed as the 'father' of operant conditioning, are sitting on your shoulder whispering sweet, significant somethings. But if your dog is not upset then the solution becomes even more straightforward. You put your operant conditioning hat on and get down to training an alternative behaviour and timing out any behaviour that you don't want. If your dog is doing something you don't like and he's not upset, then it's being reinforced somehow. It's that simple.

Of course, there are plenty of variables that need to be considered when you're constructing an effective training plan. How skilled at training is your dog already? What is he motivated by? How many competing motivators might there be? How long is he likely to train for before he loses focus? What is he capable of now? But ultimately, the answer to 'Is my dog upset?' is all you need to know to start formulating the plan.

You could be forgiven for being sceptical at this point. 'Surely', I hear you cry, 'I need to know *why* my dog is upset don't I? Why is he upset by men with beards? Why does he freak out at other dogs or people he doesn't know? Why? Why? Why?'

WHY NOT WHY?

But here's the thing. If you've got the assessment right and you've answered the question 'Is my dog upset or not?' correctly, then the route to successful behaviour change becomes clear. Upset? Change the emotion using Pavlovian classical conditioning. Not upset? Train an alternative using operant conditioning.

It's great if you have some history that accounts for the behaviour in some way. It may even help to focus on a more specific trigger or stimulus, but it doesn't change what you do. Here's an example: A dog is upset by men wearing hats. Why? I don't know. Solution: desensitisation and counter-conditioning to men wearing hats.

Now imagine if the answer to the *why* question was: 'Because when he was young, a man with a hat was very unkind to him and frightened him frequently.' Solution: desensitisation and counterconditioning to men wearing hats. Get my drift? The answer to the *why* question is only important if it changes what we do, which it rarely does.

PATIENCE, PERSEVERANCE, PRACTICE

It may sound perilously close to a political slogan that you might see on a lectern somewhere, but it isn't, I promise. It is, however, not a bad mantra to live by when it comes to working with fearful or anxious dogs. Patience is such an important part of what I do, and I try to shake a little off on to those people who, after all, are going to be doing most of the heavy lifting. In fairness, many will already have acquired a mountain of patience long before they contact me. In most cases that involve fear and anxiety, it's not going to be a quick fix. The exception is when the problem proves to be medically based and a course of medication puts it right; there's an example of this in Dora's story later in the book.

Some behaviour problems are both easier and quicker to resolve than others, but they still couldn't be characterised as 'overnight' solutions. Resource guarding, which we'll talk more about in Chapter 11, for example, can resolve well (and usually completely) with the right training. Fear, on the other hand, is an unknown quantity. You may get all the way, or you may face a much tougher mission that feels impossible at times. If you are faced with fear in your dog, then I hope that this book will encourage you that progress can be made and that you can achieve a normal life with your fearful friend. The patient work is worth it.

Perseverance is undoubtedly the most valuable quality. You either know you already possess it, or you will have to cultivate it. It doesn't need to be a relentless dedication to the exclusion of everything else in life, but steady and sustained application is going to pay dividends.

Practice? Well, you can't expect to be good at everything in life, especially if it involves technique, timing and knowledge, so cut yourself a break if it seems to be going slowly at times. You'll soon learn to follow and execute the plan and if you have a professional on board, badger them to explain the tricks. It's what they are there for. They should be supportive so if they're not then time to move on.

SHOULD HAVE, SHOULDN'T HAVE, OUGHT TO, OUGHT NOT TO

There's an important consideration that I try to hold front and centre in everything I do with dogs, both those that are fearful and those that are 'not upset'. There is no should be doing or shouldn't be doing. It's not about what they ought to do or ought not to do. It's about *what they are doing*.

Consider that for a second or two. As a society, we have a clear idea of what we expect from the animals who we either choose to use as working dogs or have as companions. We have an expectation of how life is going to be with them. They *should* be obedient. They *should* be compliant. They *shouldn't* bite us or anyone else for that matter. They *shouldn't* make a noise, except when we want them to and then only for a short time, after which quiet is the preferred state of being. They *should* welcome our guests but repel burglars, despite having little or no terms of reference for distinguishing between the two.

That's all quite a big ask, don't you think? For some dogs it is. A very big ask. And then we insist that they don't express their uncertainty or fear in the only way they know how. In the case of dogs who resource-guard that's something between a low guttural growl to a full-on assault. For dogs who fear either people or other dogs it could be anything in the flight, fight or freeze spectrum.

But when it comes down to it, if your dog won't let your partner back in the bedroom, is holding your arm to ransom over a precious toy or has the unsuspecting plumber pinned down in the kitchen, what they should or shouldn't be doing is pretty irrelevant. It's *what* they are doing that is the issue.

OF DOGS AND HUMANS, LOVE AND FEAR

Perhaps it's time to be controversial. I'm feeling sparky so let's do it.

I said at the top of this chapter that it's not surprising that people anthropomorphise their dogs' behaviour. I'll go further and say that while it's important to recognise that many aspects of our human lifestyle may not be in our dogs' best interests (our couch potato tendencies or cravings for sugar and salt, for example), I'm going to stand up for anthropomorphising dog lovers everywhere – at least to this extent. If you want to call your dog a furbaby, that's just fine by me. If you want to give them the odd bit of your chicken from dinner, or let them sleep on your bed, then I'm with you. Our dog Ripley is at the same time Peaches, Poppet, Dolly-Dog, Peanut, Pumpkin and a hundred other ridiculous names. And, yes, we give her a voice, vocalised by us so that we can have conversations with her. Go on, you've been there ... admit it!

But that doesn't stop us recognising that she is a dog and understands the world in a dog way. We need to remember that if we are to understand what influences our dogs' behaviour. That's never more important than when it comes to understanding fear.

It doesn't take long to 'install' fear in any animal, be they human or canine. But it can take a painfully long time to recover from it and that recovery can not only take a long time, but, if it's severe, will almost certainly take a village of support. We don't expect people with emotional problems to bounce back because we've told them to pull themselves together and have given them a couple of days off work. Most of us know (or at least have a sense) that it's a bigger deal than that.

'It's a dog. It shouldn't have taken this long.'

If I could assign only one phrase in the English language to the waste bin of literature, 'shouldn't have' would be it.

Our dogs deserve as much kindness, patience and empathy as any human member of our family. There's no statute of limitations on emotional health problems. All we can do is remember the fundamental principles of classical conditioning and do our best to abide by them.

So, what are those principles? The discoveries of Pavlov have paved the way for behaviour science to develop some really effective ways of changing the emotional responses that animals have to disturbing stimuli. We now appreciate the role genetics plays in the mental health of a subject, as well as the dietary choices that might be made for the prenatal mother. We understand more about the impact traumatic events experienced by the pregnant mother can and do have on her offspring. We can appreciate the importance of early imprinting and socialisation on the young.

Early experience and unexpected setbacks throughout life can have a profound influence on any animal. Never underestimate the power of fear and just how important it is that we 'pad' our own young and those of our non-human life-mates to its influence by exposing them to plenty of positive experiences, thereby developing their resilience to encounters with the new.

The basic way to think about classical or Pavlovian conditioning is that it pairs unpleasant things (such as scary people or other dogs, or people approaching your stuff) with something that is not scary but is extremely good, such as food, attention (if it's desired) or play. Food is usually, but not always, the big hitter in this regard. (We'll talk a little about what to do when food doesn't cut the mustard when we look at one of the real-world situations that will punctuate this book.)

The goal is to pair a scary thing with a good thing. A really good thing. The thing that really floats your dog's boat. Really, really good. Not just pretty good ... You've got the point right?

All well and good so far. The tricky bit, given that we're dealing with the inconvenience that dogs are living beings who may occasionally throw you a curveball, is keeping them under their fear threshold while we create plenty of those positive associations.

The fear threshold is where they exhibit signs of stress and fear when they are exposed to the scary thing. And remember, too, that we don't get to decide what is scary; they do. It might seem irrational or even ridiculous to us, but to them the world is tumbling down on top of them.

Don't think about whether they *should* be frightened or not. Accept that they *are* frightened and follow the system. This isn't a 'how to' book so you won't find a comprehensive guide to the system here, but you will get a helpful overview that will hopefully change the way you think about your dog's behaviour. My advice if you have a fearful dog, be they exhibiting aggressive behaviour or not, is to seek some well-qualified professional help.

There's a common misconception about dealing with fear-related issues in dogs: that you need to expose the dog to what triggers the fear before you can start to resolve it. The reality is the opposite. The aim is to never see the problem occur. You start where your dog is fine – not just 'kinda' fine, but absolutely relaxed and comfortable; you then expose them to the trigger in a way that they can stay feeling fine. The most common parameter for 'feeling fine' is distance from the stimulus or trigger but it could be the volume of a sound or the size of an object. Whatever that is, that is where you start. See the stimulus = get 'paid' with something delicious. When the stimulus disappears = pay stops. That order of events is important because it's crucial that the scary thing predicts the food and not the other way around. The change we are looking for is a change in the emotional response to the problem and so we need to make a predictive relationship from the bad thing to the good thing.

At some point (hopefully not too far in the future) the dog starts to show signs of anticipating the pay-off when they see the trigger. Scary things predict yummy things! Then, and only then, do you take a couple of steps forward or increase the volume in preparation for the next reveal. Although progress does not occur as a straight line on a graph (plateaus and dips are to be expected), the process does tend to speed up over time, provided the dog doesn't experience prolonged periods of stress. It's a delicate balance, but there's nothing to be gained from deciding beforehand just how long the process should take. Only the fearful animal can decide when they are no longer fearful. Our job is to go resolutely at their pace.

I lazily used the word 'decide' there, which suggests that the state of fear is a conscious choice, perhaps even designed to frustrate us or prolong the access to the chicken that's on offer. This is not the case. The truth is that fear isn't a conscious decision about anything. It's installed regardless of intention or will and the response (or responses) to it is pre-installed. It's held in the 'operating system' of our genetic make-up, to be activated in the event that a critical incident occurs, so we are ready for fight, flight or freeze. It is, in the first instance, an involuntary response to a perceived threat that the subject then learns is effective in protecting them from the imminent danger. 'Perceived' is the essence of the problem. In reality, the fear-evoking stimulus may not be dangerous but all that matters in order to make intervention by us justified, is that our dog perceives it as such.

EVERY ENCOUNTER DESERVES REWARD

Probably the biggest revelation for people when it comes to classical conditioning in behaviour modification is that as long as the scary thing is present, the dog gets paid, regardless of what they do. They can be barking and snarking, growling and spinning, but if the scary thing is there, they get paid (assuming they will eat at that point). Or you take them away to get them under the fear threshold and you pay them as soon as they'll take the food. The pay is not contingent on their behaviour. It's just contingent on the presence of the scary thing.

The usual response to this is: 'But aren't you reinforcing the fear/bad behaviour?'

If that was your reaction, I'm not surprised. It seems to make sense, doesn't it? Let me change your mind.

The crucial thing that needs to happen if classical conditioning is to be successful is for the pairing of the good and bad thing to be as reliable as possible. It's about changing the emotion and not the behaviour. The unwanted behaviour will change as the need for defence goes down. A fear response isn't motivated by the desire to get to your chicken. It's motivated by the need to increase distance between them and the scary thing. The food makes a positive connection to the scary stimulus. You can't reinforce an emotion, only change it. Reinforcement is a concept related to operant conditioning or learning by consequences, which is what we'll come to next.

LEARNING BY CONSEQUENCES (OPERANT CONDITIONING)

Think about this for a second or two. Imagine that you have a favourite bakery who bakes the finest sticky buns you have ever tasted. You usually stop off every morning as you make your way to work. You grab a coffee and one deeply cherished, sticky bun to enjoy while sitting in slow-moving traffic. On this particular morning you're running late. You could park in the car park a couple of streets down, but that's a five-minute walk and a minimum £8 fee and you can only pay using an app that you don't have and that is going to take time to download. You drive past the bakery only to find that the few free parking spaces available are all taken. The only other option is to risk parking in a 'no parking' area, but if you get ticketed, that means a £10 fine. So you are faced with a choice. Do you risk the £10 fine and save some time? After all, you may not get a ticket and then you're £8 better off than you would have been had you parked in the car park. Here are the possible scenarios:

Scenario one: You park in the no parking zone and you get away with it. You get your sticky bun and your coffee, you don't pay the £8 fee, you don't get fined and you get to work on time. Next time you go straight to the no parking zone.

Scenario two: Everything happens as in scenario one, except that you do get the £10 fine for parking in the no parking zone. But since you've got everything you wanted and you're on time for work and you've only spent £2 more in the shape of a fine than the fee you'd have had to pay to park in the car park, the gain is worth the pain. Next time you are late for work, you take the same risk of being fined.

In both of these scenarios the act of parking in the no parking zone can be said to have been reinforced. Why? Because the behaviour is repeated and so its frequency has gone up. On the profit and loss sheet of life, profits have exceeded losses and so the behaviour in question (parking in the no parking zone outside the bakery) has gone up. It's basically market economics: profit, loss, demand, supply, value, etc.

Now imagine the same scenario as above but with the risk of a £100 fine. You chance it the first time and get away with it, so you do it again (reinforcement). However, the second time you come out of the bakery to discover that your car has been towed. You have to go to the pound to collect it, pay the £100 fine and you're disastrously late for work. To add insult to injury, you even drop your sticky bun. I'm not sure what you were muttering under your breath but it's probably not good. Next time you go to the bakery you park in the car park and pay the £8 fee.

The behaviour of parking illegally outside the bakery has been punished. How do we know? Because we have a behaviour in a context and we have a measure of what happens to that behaviour: we have the same context occurring again and something happens to the behaviour – it goes down in frequency. You don't park illegally in front of the bakery again.

What drives behaviour? Consequences. Something you do causes something to happen, which is perceived to be either good or bad. If the behaviour increases, it must have been reinforced. If it goes down, it must have been punished.

In the case of this scenario the pain exceeds the gain and so you 'learn' to avoid the risk in future. The £100 could have been used to buy something enjoyable or even important. It was a costly decision.

CHAPTER 4
NATURE AND NURTURE

Like father, like son. Like mother, like daughter. Like mother, like son. A chip off the old block. Does this concept sound familiar? I imagine it does, but how much of it is true? Are we products of our parents' post-birth influence or quite literally semi-clones of them; the DNA-replicating machines described by Richard Dawkins in *The Selfish Gene*? (A nerd's read but worth it.) And what does this mean for our dogs and how they learn and behave?

It's almost inevitable that at some point in looking at dog behaviour you feel compelled to discuss the subject of nature and nurture. How much of behaviour is inherited and coded into the genes and how much is the product of upbringing and environment? As we'll see in a later chapter, dogs have been selectively bred for particular characteristics for hundreds of years, and groups of breeds are still associated with certain traits. But it's by no means that simple and when it comes to the nature-nurture argument, there's a real risk that you bounce endlessly backwards and forwards between the two. My answer is 'both', but that just doesn't seem adequate. If anything, behaviour is something of a love child of the two, which is sweet isn't it?

Why? Well, here's my case. What is DNA for and what do genes do? They are all of the information and tools for cells to build the various components or proteins of a human body. Our DNA and genes 'decide' what specialism each protein will have. Specialists in eye construction, therefore, need a blueprint to tell them what colour to make those eyes.

There is, without doubt, a bundle of genes that mean we, as newly built humans, are predisposed to look a certain way. We are also pre-programmed with a pack of behaviours that will pop out as we grow or are ready to rock and roll, fresh out of the showroom as it were. It stands to reason that we have genes that govern our ability to learn. Learning is going to be really important as life goes on, not least because we need to make it to reproductive age without anything grim happening to us. The ability to learn through experience (nurture) is pre-installed in the software (nature). The perfect love match.

LEARNING IS IN THE DNA

All of which is fundamental when we think about nurture and nature and the effect on dogs and their behaviour. There have been two recent research studies that, though they aren't exactly in completely harmonious agreement, do at least concur on one aspect: trainability is inherited. It's part of a dog's DNA.

A 2019 study published by The Royal Society, using a sample of 14,000 dogs, looked at the heritability of behaviour traits in different breeds of dogs.[1] The traits with the highest breed heritability were: trainability, stranger aggression, chasing, and attachment and attention-seeking. These aren't altogether surprising, given these are the exact kinds of behaviours humans will have selectively bred for in their dogs over hundreds and hundreds of years. Stranger aggression might serve to protect flocks and alert to intruders. Chasing would help in both hunting and (down the line) herding breeds. Attachment and attention-seeking might be accounted for by the shift in importance during selection for companion-appropriate behaviours. But the gold ticket item here is 'trainability'. Jackpot. Trainability is an inherited trait expressed by genes (or a collection of genes) in the brain.

Another study by experts in dog genomics, used data provided by community science project Darwin's Ark. This study surveyed 18,385 dog owners and then genome sequenced 2,155 dogs.[2] The study came to the conclusion that dog breed served as a poor predictor of behavioural traits in any individual dog. Breeds that are still most often bred for working traits (e.g. Border Collies) are more likely to maintain a strong correlation, but on average, breed accounted for only about 9 per cent of the variations in any given dog's behaviour.

That doesn't mean some breeds are not more inclined to show some behaviours more readily than others. These fixed action patterns (or innate behaviours) are 'pre-installed' and, arguably, still genetically present in all dogs. If we were to travel back in time to those early wolf-dogs, they would likely exhibit all of them more obviously than modern domesticated dogs. Modern dogs exhibit some of those original, pre-installed behaviours more than others because of their physical ability to do so. (Think of how Border Collies are physically adapted to chase and herd because of their athleticism. French Bulldogs wouldn't have a hope in hell of rounding up sheep with the same efficiency.) The more we humans have selectively bred dogs to do specific jobs, the more they have changed physically which, in turn, has made some behaviours more likely.

It might seem counterintuitive to say then that breed is a poor predictor of behaviour. But what I think the authors of the study are saying is that *some* really well-entrenched stuff can be relied upon (although not universally – not all Border Collies are good at herding, not all Labradors will become efficient at retrieving and not all German Shepherds will be any good at chasing down bad guys). However, if you are going to hang your hat on breed for anything less task-related (like 'friendly' or 'good with kids') you could end up either surprised or disappointed. Perceptions that one breed is more or less likely to present with one behaviour problem or another are exactly that: perceptions. The data for maintaining otherwise is poor, but a dog's physical ability to 'go big' with a particular behaviour will skew people's view.

In short, you're just as likely to get bitten by a Yorkshire Terrier for taking their bone away as you are by a Dobermann.

But the big message, if this research proves to be right, is that breed doesn't serve as nearly good enough an indicator of behaviour than we have previously been led to believe. If it is right (and it is a thorough piece of research), then that throws the whole issue of breed-specific legislation (breed bans) in the air.

Here's the thing though – regardless of the impact of breed on behaviours, both studies agreed that trainability is inherited. In fact, wherever you look in reputable journals, you find reference to the very high heritability of trainability (or biddability, depending on what you read). It's almost as if nature, as beautiful as it is, is acknowledging its flaws, as if it's saying: 'I don't always get it right, you know. I make mistakes. Sometimes this remarkable genetic experiment throws in a curveball, a bug in the code, if you like. But, hey! Don't worry. I've built in a failsafe. I've built in a "trainability" gene so you can repair the system should you need to.'

Which brings us on to epigenetics. I confess to being fascinated by this subject and it has become the buzzword of the past few years. As far as I understand it, epigenetics is the potential to manipulate gene expression, not just through actual physical changes to DNA, but by environmental interventions too. That suggests that mere mortals like you or I can alter genetic expression through what we do and the decisions we make. We can not only teach dogs to do things differently, but that can have an effect all the way down to the genetic level. It affects the very DNA by turning genes on and off, a bit like light switches ... I think!

Genetic make-up, so we are learning, is much more 'plastic' than we might have imagined, which is a good thing in my view. It rules out genetics as an argument used by poor training practitioners for behaviour being 'unmodifiable'. Instead, it implies that all those things we once believed to be immutable are not necessarily so. 'Sorry, but we're fighting genetics here.' If you've ever heard this from a trainer, perhaps because of the particular breed of your dog or because they are displaying aggressive behaviour, then whatever polite reply you come up with, let your secret self be thinking, 'Oh really? Is that the best you've got?'

If I'm feeling generous I can see where they're coming from. We've long been told that your genes are your genes and it's futile going to war with genetic hardwiring. But that's a get-out-of-jail-free card, isn't it? 'I'm afraid your dog has the psychopath gene, sir. We're going to need *Silence of the Lambs* level security in place and then we'll see what we can do about improving his recall.' OK, that is a little flippant of me. Some problems are so insurmountable, and the level of risk so great, that behaviour modification is too difficult to do safely. But I don't think we should be labelling all dogs, in particular those who exhibit aggressive behaviour, as beyond help because we either aren't experienced enough or knowledgeable enough to provide that help.

It also means that as dog-loving people, we should never assume that our dog absolutely *must* behave in a certain way because of their breed and that there's nothing to be done about it. Even if there is some historic or prehistoric disposition, it's still all to play for in training terms.

CHAPTER 5

WHAT ON EARTH ARE YOU TALKING ABOUT?

'Research has shown that dogs and cats can feel happy or sad, excited or disappointed, depressed or elated, and that these feelings may be similar to ours, even if they cannot express them in the same way that we do.'

– Noel Fitzpatrick, *Listening to the Animals: Becoming the Supervet*[1]

We're a funny species, humans. We expect our dogs to pick up our language and listen to us rabbiting on about work, the weather or the state of the economy, but not many of us make too much of an effort to learn theirs. In some ways, that's fair enough: there isn't a helpful online canine dictionary or grammar primer and despite the richness and eloquence of our very own dog's different modes of communicating, we won't be seeing a Ripley's Thesaurus on the shelves anytime soon. Fortunately, our understanding has come a long way over the past few decades, and we know much more about how they communicate with each other and how they try to make themselves understood by us. But there's still much to learn.

Think of this chapter as a kind of 'Get by in Dog'. This isn't about UN interpreter level fluency — no one could claim that. But I'm going to try and give you some tips on mastering the basics.

BREAKING DOWN COMMUNICATIONS

There is no getting away from it, communication is a tricky road to travel. Let's imagine a meeting of a group of people. They live in the same country, but not in the same region. They all share the same first language. So, with all that in mind, you would imagine that communication between them would be straightforward, wouldn't you? No? Really? You're on to me, aren't you? What about slang? Each generation creates its own vocabulary, which is, more or less, incomprehensible to the one that went before.

And what about dialect and accent? Nina and I moved from the south-west of England to the north-east of Scotland. We'd spent decades getting used to the language of Cornish pasties and King Arthur to voluntarily up sticks and travel to the north-east of Scotland, the land of Doric. Don't get me wrong, I love dialects and accents and we love living up here in the tip of the country. We did buy a Doric dictionary in the hope it would assist in translation (it didn't), and we are continually charmed by the willingness and endless patience the people here show in repeating things slowly for us. Wonderful.

Now imagine a group of people who do not share the same mother tongue. Some might know the basics of a second or third language that some of the others also understand. That makes it easy enough to get by in communicating, provided things don't get complicated.

And if they do get complicated? The potential for misunderstanding and misinterpretation grows. If the purpose of the conversation isn't serious, then that probably doesn't matter too much. But if making sure you don't offend anyone is paramount, then things just might get delicate.

In my scenario here, all of the participants are people, and people have ways of understanding other languages. There are conventions and – sometimes – shared structures. And, if all else fails, there's Google Translate! But what if one of the group is from another planet? It's their first time holidaying on Planet Earth and they left their handy Babel fish translator at home. Now things get really tricky. No common structures, no vocabulary that could be easily identified, not even an alphabet, as far as anyone else in the room can tell. The prospect of a serious misunderstanding just grew exponentially. That is pretty much the scenario you have when you put two different species from Planet Earth in the same room. And if one has big teeth, it's quite important to open a reliable channel of communication.

FACE TO FACE

The first thing to say is that most of dogs' 'conversation' isn't vocal at all. (Do I hear disagreement at the back? We'll come on to barking later.) Dogs largely use body language.

We human beings are not naturally great at this form of dog-to-human communication. Oh, would that we were, but straight out of the box we are really not good at it at all, which is surprising given how important it should be to us. The failure of humans through the ages to interpret dog body language efficiently and accurately could have significant implications, particularly in potentially dangerous situations, so given how long we've been knocking around with dogs, it's curious that we haven't got an awful lot better at it.

Our inability to read dog facial expressions is a case in point. A 2020 study at Lincoln and Leipzig Universities found that we were particularly weak in this department.[2] As far as human faces are concerned, we're hardwired to scan them in a particular order and then zoom in on the area that is going to give us the most information, having established the general emotional situation. We tend to look at eyes in negative situations and the mouth in happy ones.

However, dogs' faces express things differently and so you would imagine humans would at least settle on which are the more telling areas, much in the same way as we do with one another. But not so. We scan dogs' faces in the same way as we do each other's and then focus on the same locations as we do for humans. Usually the first scan is so hopeless that there is too little information to help a more valuable analysis. For example, we are inclined to completely disregard the ears, both in the first scan and the subsequent detail gathering, and yet ears are an important part of a dog's expressive repertoire – dog ear positions are multi-purposed – but other gestures and context should be considered as well. But, without training ourselves to do so, we're really bad at observing ear positions; in fact, most of us need to stay behind after class and do an awful lot more work to be able to come anywhere close to interpreting what our dogs' faces are telling us.

BODY LANGUAGE

Nevertheless, there are some elements in our cross-species conversations that have become intrinsic and almost instinctive. Lots of our dogs' body language we find easy to interpret. There's little doubt what a wildly wagging tail, pulling your dog's rear this way and that, means, despite no one having told us. Most of us would just get it. But there's plenty about dog behaviour and their attempt to telegraph information to us that is all too easily misunderstood. Often the misunderstanding is harmless. The rules of a ball game with Ripley can be hard to fathom, for example. She offers the ball, but sometimes, it seems, the way we throw it is unsatisfactory and so we're penalised – the ball is snatched back before we can pick it up and no further throwing opportunities are offered until we've spent some time in the sin bin, thinking about our behaviour. Ripley's Rules Ball is far too complex to be communicated to humans with their limited linguistic abilities, so we just have to muddle through.

Some of our dogs' valiant attempts to communicate with us, however, have more serious implications. It's helpful if we don't misidentify important signals like 'Stop', 'Ouch', 'Go away', 'That's scary', or 'Bobby's stuck down the well!' If we misinterpret 'That's scary' as belligerence, it can result in a serious breakdown of interspecies relations.

Now, we're not all linguists. My GCSE French teacher once said to me that he didn't care what I did in class, as long as I didn't disturb anyone else. I had a natural propensity for other academic subjects, but modern languages were most certainly not among them. Fortunately, dog body language isn't littered with prepositions or gender-specific nouns. You won't need to learn that chairs in French are feminine, but masculine in German. You won't have to sweat over complex sentence structures or obscure rules of punctuation. But if an angry and misunderstood dog is holding your arm as insurance, you won't be able to resort to Google Translate either, so it's helpful to be able to read the signs with some accuracy.

There are some challenges we need to acknowledge at the outset. It's not an exact science for one thing. We can't ask a dog what their intention was or is regarding a particular behaviour. Observing what a dog does and describing what we see is not the same as knowing with absolute certainty what that behaviour means. Interpretations of what dogs do, particularly if based on a still photograph or a few seconds of video, can vary wildly and the confidence with which those views are held can be disproportionate to the evidence available.

UNDERSTANDING THE BIGGER PICTURE

There is often a fair amount of bias and intuition involved in our interpretation of dog body language and I'm not going to dismiss that out of hand. It often has validity, particularly if the interpretation is made by people with a lot of observational experience who have taken the time to track and match behaviours with outcomes. Some things are simply easier to track. A play bow, which we will look at more closely later in the book, is an obvious signal with an obvious intention when you see how often the outcome is play. You can count how many times you see it and you can tell who it is directed at; you can get data from it. But that little lick lip they just did? Did that mean something, and if so, what? Can we accurately say what the antecedent was, or if the consequence we've assigned to it is right or wrong?

Does it signal intent or is it an instruction? Should we stop what we are doing or is it intended to signal a wish to proceed in a friendly way?

With that in mind, I'm going to admit that some of what this section holds will be informed opinion on my part (or bias if you prefer). This is simply because there is less conclusive research available on that particular subject. Other aspects have some really solid scientific evidence behind them and I'm always going to go with that. As far as I'm concerned, anything else would be sacrilege.

One thing is for sure: when it comes to how dogs use body language to communicate, having a substantial amount of observation under your belt tends to lead to a reluctance to make swift judgements based on snapshots and a greater inclination to look at the bigger picture – and that is always important.

The need to see the bigger picture can't be emphasised enough. Snapshots and short videos are interesting, and you will often get plenty of opinion on them, particularly if they are posted on social media. But what images *don't* tell you is what happened before or after, or what might be happening outside the frame of the photograph or video.

You might be forgiven for thinking that a 10-second video clip would give you more than enough to judge what's going on. It must be obvious, surely, if a dog is upset by an encounter or not, right? Perhaps, but not always. If you have extended knowledge of the dog, then you will have more insight. It goes without saying that if you are part of the dog's official care team, or have been given a detailed background history and been invited to contribute, then you are going to have quite a bit to go on; there's an excellent chance, therefore, that the collaboration will yield some solid understanding.

Observations from uninvolved commenters can sometimes be useful, so I try not to dismiss them out of hand. They might give you pause to re-evaluate or check back with your client about, for example, the sequence of events. I particularly like it when comments on social media start with a question: 'Could that mean …?' 'I've heard that …' 'Does that apply here, do you think?' It's also good to acknowledge limitations: 'He appears to be stressed, but what is he stressed about? Is it that attention is unwanted or is something else going on?' (I really like that last one.) But judgements based solely on what is essentially just a tiny slice of an event without any real context are, as often as not, going to be wrong.

Here's an example. You're scrolling through your social media feed and there's a 10-second video of a dog being stroked by someone familiar to them. The dog's ears are pinned back against their head. Their head seems turned away from the person, but their eyes are turned back towards them. Their back is a little hunched and their tail is tucked. You see two or three tongue flicks in the space of that ten seconds.

So what do we have? Should the familiar person be stroking the dog? What do you imagine has occurred before the video sequence you are watching? What happens next? Does the dog bite the person? That's a lot of questions, isn't it? And you probably feel that the answers could be drawn from the video. That could be true. You could guess right. But even if your guess is an educated guess, it would be a guess all the same. Photographs and videos make excellent tools for helping people with resolving behaviour problems, but context is everything. Once you have context, you have the beginnings of a conversation that will lead to greater understanding and, ultimately, a solution to the problem.

Let's get back to our unhappy-looking dog: ears plastered back, hunched posture, tail tucked, lip flicks. That last one is ambiguous in my view, but collected together with the rest, it doesn't make for a good picture, for sure. But what is the picture of? I would guess that most people would assume that the dog has been called over or approached by the human. There may be a punishment history behind the posture and appeasement gestures, or there might be something about to happen that the dog doesn't like, such as grooming. The dog, you assume, is now forced to endure physical contact while in a stressed state and is upset at the encounter. You speculate that two seconds after the clip was taken, the dog could well have bitten the person, been told off and started the whole cycle again. You're just a heartbeat away from getting into full-on keyboard warrior mode and making a comment on the thread, telling the ignorant individual exactly how much they're getting wrong in their handling of that dog. But let's just take a breath here. You could be right. You've certainly answered the all-important question correctly. The dog *is* upset. But did you ask the follow up question? At *what?*

Here's what really happened. The dog was in another room while the person (the one visible in the video) was in the kitchen. The house next door is having some renovations done and there is scaffolding being erected at the front of their house. The dog is afraid of unexpected, loud noises (pretty much anything sudden and percussive). One of the workers next door drops a length of scaffolding on to the lorry and it collides with the scaffolding already there. Most of you will be able to imagine what that sounds like.

The dog runs into the kitchen from the other room and takes a hiding position on the other side of his trusted person, turning away from the threat and looking back at his human. He even looks up at them for a moment. Knowing what came first (the antecedent), what happens (the behaviour) and what happens next (the consequence), you can make a pretty solid evaluation of the dog's motivation. The dropping of the scaffolding has frightened him. He is responding fearfully to the antecedent and is seeking out protection and reassurance from a valued secure base: his person. This person knows the dog is easily frightened by loud noises and, based on experience, has confidence that physical attention helps their dog. They are sure of this because they do regular consent tests (the practice of stopping the affection to see if the dog requests that they continue).

So the bigger picture tells a very different story from what you might have imagined.

That was a very nuanced set of behaviours. Some of the more straightforward examples I'm going to come on to here will, I'm sure, be familiar to many of you from your observations of your own dogs: how they act with other dogs, familiar and unfamiliar people and so on. You've most likely got a mental checklist of their body language that you feel confident about.

UNDERSTANDING THE ESSENTIALS

So let's take a look at the dog body language that's pretty essential for most of us. Given that my speciality is fear and aggression, we'll start with the big-ticket items: what tells you that a dog is friendly and what might tell you that they are not. The first thing to say, and I would shout it from the mountain tops given the opportunity, is never, ever, approach a dog you do not know, however confident you are of your own experience with dogs. Often a dog might telegraph some very obvious prosocial signals, but as a rule I would encourage anyone to wait for the dog to approach you, signalling peaceful intention, rather than for you to march in there, assuming your overtures will be welcomed.

WHAT ON EARTH ARE YOU TALKING ABOUT?

It's like that scene in the sci-fi thriller *Independence Day*, where the US government 'Welcome Wagon' helicopter approaches the alien spacecraft, lights flashing to explain they come in peace, only to be blasted out of the sky. Not a good outcome. Best not to assume your peace mission is shared.

Right, let's lighten things up. The dog in question has approached you. You haven't done much at this point to allow the investigation of your person to commence, so what should you expect? Wagging tail is usually a good sign, particularly if the tail is set midway or parallel to the floor and the wagging is either wide or rapid. Dogs who haven't been trained not to jump up at visitors, may jump up and attempt to reach and lick your face. It's how they greet one another, so it should not be too surprising that they try and do it with us. For those dogs who have been taught not to jump up, they may well approach while tongue-flicking to mimic the greeting ritual. Prosocial dogs will often plaster their ears back (something that is often associated with anxiety, but that's a good example of how important it is to pay heed to context). They might paw at you or raise their paws to you and then there may be a whole range of vocal noises from whining to barking or even growling/grumbling.

So far so good! These are some hints that an interaction with a dog is likely to go well. But what about those clues that suggest leaving 'stage left' might be the prudent option. We have already talked about the most obvious one:

if they don't come to you, then I would, frankly, take the hint. It could be that they are just doing a bit of a Greta Garbo, but even without the dark glasses and headscarf, keeping a distance is a strong signal of 'I want to be alone, darling'. The old-school idea might have been that dogs should do what we want, when we want them to do it and that they should jolly well tolerate it. But things have moved on and now mutual consent is the name of the game. It is a much safer place from which to start a healthy relationship, however fleeting or casual that relationship might be.

The more obvious explicit indicators that a dog *doesn't* want to be your friend are commonly known as threat signals. They are designed to get someone or something that is unwanted to move away or stay away. Some are unmistakable; as humans, we seem to be pre-programmed to get the point, even if the decisions we make as a consequence are not always the smartest. Dogs give us tip-offs that they would like us to go away; they have threat signals they can deploy to affect that. Snarling, hunkering, a 'hard eye', making themselves tall (sometimes in conjunction with stiff movements and their hackles up) are some of those you might witness if you get too close to them or their valued resources. These threat signals will sometimes be displayed in conjunction with other signals that are widely accepted to be appeasement behaviours and that indicate fear. These dogs are ambivalent. They might show some signals that are designed to make themselves seem smaller and so less threatening, while at the same time issuing threat signals, which need the recipient to perceive them as dangerous to have effect. These dogs are conflicted about what strategy they should employ to deal with the danger. Whether it makes them less or more dangerous is open for debate.

While we're on the subject of signalling, it's popular among some dog trainers to see behaviours like yawning, lip-licking or turning away as calming signals. It's an idea that came to the fore with the publication of Turid Rugaas's book *On Talking Terms with Dogs: Calming Signals*.[3]

Some behaviours attributed as calming or appeasement are difficult to be entirely confident about. The more cautious observer would point out that some of these 'calming signals' are displayed even when the dog is alone and there appears to be no one around to receive them. In which case, what has motivated the behaviour? Is it a noise, the source of which is out of sight? It could mean that the behaviour serves to prepare the dog for whatever comes next, readying for something. Who knows for sure? Some lean towards interpreting them as a physical stress reliever, like squeezing a stress ball, rather than a signal intended for another. And some in the dog world can become really quite adamant that behaviours like lip-licking are always stress relievers and clear indicators of a dog in distress. But this is not necessarily so. As always, context is all. What's the environment, what's going on, who else is around and are they familiar or unfamiliar? To my mind, there is no reason why the same behaviour could not serve both purposes. And I tend to the view that when they are used as signals, rather than being intended to calm the receiver, they are transmitted to signal a desire for a peaceful encounter on the part of the sender: 'I would like a peaceful meet and greet. What would you like to do?' Again I can't be sure of this but it's worth considering so that we don't get over-cautious when it comes to our dogs meeting other unfamiliar dogs. They might not be stressed, just cautious, and making a first overture to say: 'Do you want to be friends?' Still, there's room for quite a bit more research on this subject.

Before we leave the territory of behavioural signals, there's one more we need to look at. You may very well have seen it in your own dog. It's the head tilt and dogs often do it when you're talking directly to them. They'll hold your gaze and tip their head to one side and then the other. What's that all about? Are they exploring your face to divine your intention and listening intently for a familiar word that might give them a clue – something good like 'biscuit' or 'dinner' or 'walk'? Are they stunned at the absolute nonsense you're talking about politics, or who is going to win *Strictly* and tipping their head in disbelief? Or are they drinking in your every word, in awe of your intellect, pleased, proud and privileged to be the dog of a towering genius? Well, the latter, obviously. No question.

TALKING THROUGH PLAY

I am a huge fan of dogs playing. Who isn't? Anyone …? I didn't think so. Good. In that case let's talk about how dogs talk to each other in play. How they keep things within acceptable boundaries so that everyone lives to play another day.

> 'Play is the highest form of research'
> – Albert Einstein

You can always depend on Albert to sum things up succinctly. Play is a very versatile set of behaviours for any animal. On one level, play is life rehearsal. It's a safe place in which to practise fighting, fleeing, freezing, chasing/hunting, mating, etc. without having to go headfirst into real life with little notion of how to navigate it all. Of course, your dog isn't going into any of it with the idea of life rehearsal on their mind … or at least I don't imagine they are. It seems reasonable to suppose that evolution has developed this as the perfect way of establishing learning scenarios without the easily bored target being aware. In other words, the desire to have fun is hardwired into the brain. Clever old Albert clocked on.

We'll come on to how important play is for building relationships and creating a sense of wellbeing and security in

Chapter 8. For now, let's focus on what play does for the communications game.

It matters that puppies and dogs can interpret the intentions of their playmate accurately so that everyone plays safe and plays nice. And to that end it matters that we, as supervisors and referees, interpret it accurately too. Failure to do so may result in mishap or in the unnecessary and premature cessation of play. Boo!

There are some behaviours that dogs exhibit in play that can help us interpret what they're 'saying' too, so it is worth looking at them here. They are known as meta signals and are used by dogs firstly to facilitate play, then to demonstrate that the intention is play and finally to keep play going. Why might these signals be necessary? The clue is in the core purpose of play: life skills rehearsals. Many of these life skills include violent actions, such as fighting or hunting, and they include many of the threat signals we discussed earlier, only this time they are 'mock performances'. It's important that everyone recognises what is going on so that there are no misunderstandings. It would not do for the alien arriving on Earth for the first time to misconstrue a well-meaning gesture of fun from us as an out-and-out declaration of war. Before we knew it, there would be a flotilla of heavily ray gun-armed alien battle cruisers sitting off Ursa Minor waiting to obliterate every square millimetre of the planet. Oops!

THE PLAY BOW

Most people who have seen their dog play will recognise the play bow. It can occur throughout the play session but is frequently seen at the beginning. It is presented by at least one dog, but may be 'returned' by the recipient, presumably to indicate acceptance and approval.

The play bow is identified by the spreading of the front legs causing the head to go low to the ground. The rear end stays upright and so is thrust into the air. At the opening of play, it follows mutual investigation, which ends with a brief moment of stillness. This can often leave us humans holding our breath as the dog's body stiffens and grows tall. Is this going to be a spat? No! There's the bow and moments later a full play explosion occurs.

THE PLAY FACE

There is a lot about play that is big and rough. Body slamming, pinning your opponent and biting are all common. Even if they are not delivered at full thrust, they are nonetheless remarkably similar to the real thing and so dogs and puppies have some signals they can send that say: 'What is about to happen is not for real. I'm just kidding with you. I'm playing.' The play bow is one such signal; another is the play face. It is big and exaggerated and often precedes a play bite. The message is: 'Here I come, all ferocious and stuff ... gruffle ... Lol!'

THE WONDERFUL THING ABOUT TIGGERS ...

Big, exaggerated, over-the-top bouncy gaits are a huge tip-off that a dog intends to play. In fact, 'intention signals' is a good name for these behaviours, which indicate that the intention is play and isn't hostile. Dogs who are chasing prey or looking to get something or someone to leave because they are threatening the dog's safety or stuff, tend to run fast and low to the ground. It's a highly efficient, very flat, running style. All the dog's energy is put into the speed. Bouncy gaits are the exact opposite. They are inefficient, exaggerated and, indeed, silly. The message? It could be 'Look. This is fun. Come and join me.' Or perhaps 'Come on. Let's see who can win the Ministry of Silly Walks competition.' Or 'Chase me. Chase me.' Whatever it is, it certainly is not menacing.

THERE'S MORE TO IT

Of course, there's more to it than just three intention or meta signals, which is just as well. In the heat of a lively play session, it could start to get difficult to spot the good intention from the getting out of control. But they do have more. Paw raises, lip licks or tongue flicks, play pauses, role reversals (you chase me then I'll chase you, then you pin me and I'll pin you). Ears up and rotated back. All of these are evidence that the play is going well and is being mutually enjoyed.

There is one signal in particular that I want to mention and matters enormously if you have a gentle player coming up against a robust one or there's a mismatch in size. It's called self-handicapping and it's really cool.

'OK! I'LL BE BLINDFOLDED AND WITH ONE PAW TIED BEHIND MY BACK ...'

Self-handicapping is pretty much as it sounds. The more seasoned, larger or more robust player moderates their play to match that of their playmate. If you think about it, this is in everyone's interest. If the bolder player comes on too strong, the play stops because the more 'timid' or smaller dog feels uncomfortable or unsafe.

My favourite example of self-handicapping is during wrestling matches or play fights. The more rigorous player may go to bite their opponent, but they never close their mouths. Indeed, they have their mouth open to an almost absurd degree and wave their heads about in the most comical way.

The brilliant thing is that they'll use those signals to encourage us to play too, and as their best friends, this is a vocabulary that we should study and celebrate, too.

MORE THAN JUST A DOG

NOW YOU'RE TALKING

Of course, dogs do communicate vocally, too. First of all, they bark and the reason they bark is to communicate with us. As we explored in Chapter 2, humans and dogs evolved together, cooperating and cohabiting for mutual benefit, so it's perhaps not surprising that those clever dogs developed a language to try and tell us important stuff, such as 'there's a lion creeping across the plain' or 'the tribe from across the valley are trying to steal your spears'. Dogs might sometimes bark at other dogs, particularly if separated by a fence, but it's not their prime method of communication with their own species. And dogs' distant relatives, wolves, hardly bark at all.[*] They howl to let other wolves know where they are – a form of prehistoric GPS – but they don't bark.

Dogs bark to get our attention, to warn us about things, to signal distress or to tell us to back off. And as I touched on in Chapter 2 when looking at our conjoined evolutionary path, we humans seem to be hardwired to know at least something of what they're trying to say. Even non-dog owners can tell pretty reliably what's upset and what's hostile.

High-pitched barking rising in tone and ending with almost a howl rends the heart because it speaks of loneliness. Low barks with little space between them mean 'back off right now'; repeated and high-pitched can mean excitement and readiness for a game.

So ponder this the next time your dog is barking up a storm as the Amazon driver walks up the path. Your dog has likely evolved this bark for your benefit. If you shout at him to shut up, he just thinks you're talking back.

GROWLING IS GOOD

I have a question for you (you can keep the answer to yourself; I know this can be a sensitive topic). Have you ever told your dog off for growling? At the neighbours, visitors, the postie, family members or friends?

By way of reassurance, let me tell you that the most common answer to this question is likely to be 'yes', so don't feel judged if that includes you. I am, however, going to explain why growling is good. That probably surprises some of you; others will think I've gone all 'life-change guru'. But here's the thing, growling really is good and we as a society need to get better at acknowledging that. Punishing out the growling is a bit like a messenger telling you they have news of something terrible that is about to happen and where best to go right now, and you kill them before waiting to find out the details. Head meet sand. Stay with me now. Don't kill the messenger.

Growling is a message. Along with barking, it's part of the relatively limited vocabulary through which dogs can communicate with us. The message that growling is sending us is: 'I'm not comfortable with this situation. Back off,

Buster.' Sometimes growling is done in play and I acknowledge that, but for now we are going to focus on the 'I'm upset' intended message. I'm sure that the majority of us can tell the 'back off' growl from the 'hey, let's wrestle' one. (Although it is sometimes the case that playfulness is mistaken for hostility, but we'll get to that.)

What happens, given all we now know about how dogs learn, if you punish them for growling? If it has had any impact, he is going to stop growling. Growling predicts bad things happening. So, he reasons, stop growling and bad things don't happen. Right?

Here's the next question: Have you solved the problem? That depends on your intention. If all you wanted to do was stop the growling then well done, you've achieved that. But what if the reason you were worried about the growling was because you thought he was being aggressive and you wanted to reduce the risk that your dog will bite the next person they meet? Weeeeeell, not so much! Sorry to be the one to break the news, but you have not decreased the chance that your dog will bite people he doesn't like (and, by that, I mean people he is fearful of) one iota. Only now it's highly likely that he isn't going to signal his intention, giving you the opportunity to avert the crisis by getting him the £$%^ out of there. You see growling, in this context, is the first in a series of protracted warnings that a dog can use to communicate his discomfort. It's saying 'Ooooh! I don't like this.' By punishing out the growl, you cut off all the means a dog can deploy, with escalating intensity, to indicate he is upset, before he resorts to an all-out assault. Others in the series could be a snarl accompanied by a sharp bark, or a rapid bark like an assault rifle. I would imagine that most of us, if faced with the sound of an assault rifle, would back off.

If, by punishing growling, you cancel the sequence of protracted warnings, then the only alternative the dog has (even though this represents a personal risk) is to bite. And there's a further risk: if the dog has been punished for growling at a scary person (or scary child) to get them to keep their distance, scary people now predict an unpleasant consequence. The dog's attitude towards strange people has almost certainly worsened. Many will talk about a dog bite being 'unprovoked' or coming 'out of the blue'. But here's the thing: there's no such thing as unprovoked behaviour. You just didn't notice what provoked it. It only came out of nowhere because you killed the messenger. Growling is good!

HERE ENDETH THE FIRST LESSON

This chapter has really been just a toe-dip into understanding how our dogs communicate with each other and with us. No amount of theory and book-learning alone will make you an expert. Like learning any language, the best route to fluency is to get up close to the natives, observe them keenly and listen intently. Most importantly, keep track of the context: what came before the behaviour and what came after, was the dog in a situation that was familiar or new? That is the way towards your UN interpreter badge in Dog as it is spoken and behaved.

CHAPTER 6

BUSTING THE MYTHS

So what is the fallout for dogs of the Lassie myth? As soon as you bestow intelligence and morality, you bestow the responsibility that goes along with them. In other words, if the dog knows it's wrong to destroy furniture yet deliberately and maliciously does it, remembers the wrong he did and feels guilt, it feels like he merits a punishment, doesn't it? That's just what dogs have been getting – a lot of punishment. We set them up for all kinds of punishment by overestimating their ability to think. Interestingly, it's the 'cold' behaviorist model that ends up giving dogs a much better crack at meeting the demands we make of them. The myth gives problems to dogs they cannot solve and then punishes them for failing.

– Jean Donaldson, *The Culture Clash*[1]

There are so many myths surrounding the human–dog relationship. Despite the huge strides made in recent decades in understanding how our dogs behave and learn, they have remained remarkably 'sticky'.

Perhaps they've hung around for so long because they reflect a particular human view of the world: our experiences, prejudices and insecurities. Or maybe it's just because they're so seductively simple. They get passed on with complete conviction in puppy training classes. They're trotted out in newspaper articles and social media posts. They've had all too regular outings in dog training TV programmes over the years, making them seem unassailable, long after they've been debunked.

The sad thing is that these myths are at best unhelpful and at worst damaging to a healthy, happy relationship with our dogs. So let's do some busting. If you have held on to some of these myths, there's no shame. But it's time to let go.

MYTH 1: YOU MUST BE LEADER OF THE PACK AND SHOW YOUR DOG WHO'S BOSS

Let's tackle the big one first. I touched, in a *previous* chapter, on how our dogs almost certainly descended from a species of wolf many millennia ago. But the fact is they're not wolves now and don't act like them, any more than we, as humans, generally behave like chimpanzees with whom we share a substantial chunk of our DNA.

Despite this, the idea of the 'alpha' or the dominant leader of the pack, which started with wolf researchers has been embraced with enthusiasm in some parts of the dog world and used as a basis for defining how we should approach our relationship with our dogs. The original wolf research was carried out by Rudolph Schenkel in the 1940s and was followed by further work in the 1960s by Dave Mech. Both observed groups of unrelated wolves held in captivity and they concluded that each pack was dominated by an 'alpha wolf' who fought to gain leadership of the pack and subdued the other wolves. Even at this stage, there was no suggestion that this could, or should, apply to domestic dogs. But that didn't stop the idea gaining traction that, as dog owners, we have to be the 'alpha' in our household pack or disaster will befall us. If we don't dominate our dogs, so the argument runs, then they will dominate us and take control.

There are some clear problems with this: the first being the concept of the 'alpha wolf' itself and the second, its application to domestic dogs. Dave Mech revised his views of wolf behaviour in the 1980s after observing wolves in the wild, rather than in captivity. He found that pack leadership wasn't established through a battle for dominance; instead, wolves lived in family groups. Rather than 'alpha male' and 'alpha female' reaching the top of the hierarchy by subjugating the rest, packs of wolves were made up of offspring, led by their parents. He asked for his early book to be withdrawn from publication and reiterated many times that while he had for years studied wolves, he wasn't drawing conclusions about domesticated dogs.

And if we focus on our dogs, living with us in our homes, then the original pack dominance argument seems even more unconvincing. With our dogs, we have no need to engage in a battle for mastery of the household because we control the resources. We have the opposable digits (useful for opening food packaging and bags of treats), the bank cards (no need for all that exhausting hunting when you can just go to the shop), the keys to the home (safe, warm and furnished with comfortable places to sleep). Dogs, evolutionarily developed as they are to adapt their behaviour to what benefits them, generally know what's in their best interests. If they overthrow us, how on earth are they ever going to open the fridge to get a little nibble of cheese? Some dogs, of course, guard food, toys or locations, but this is by no means the way of all dogs and isn't about dominance. It's also easily modified through careful and gentle training. (More of this in Chapter 10 on The Fearful Dog.)

Despite all the debunking there has been over decades, pack dominance thinking has maintained a foothold. It has been used to justify aversive training techniques: physical punishment, 'alpha rolls' and aggressive lead jerks, which can cause behaviour problems and do nothing to foster a constructive bond. And it has led to a whole host of 'mini myths' around how humans should behave with their dogs, which, though largely harmless, are often inconvenient and always irrelevant. These include: you must eat before your dogs to show you are the dominant one; never let your dog go through a door first or they'll think they're the pack leader; never allow your dog on to the furniture – they should never be on the same level as you or, perish the thought, higher. If you want to keep your dog off the sofa to keep it mud- or hair-free, that's up to you, but showing your alpha status has nothing to do with it.

And if you have a fearful dog and anyone tries to tell you that this is because you're not showing 'strong leadership', you can ignore that, too. Our dogs aren't looking to us to lead the hunt or go to war. You don't need to be strutting, shouting and threatening like a dictator.

So what are dogs looking to us for? Does anything about the idea of showing your dog 'leadership' make sense? In some ways, perhaps, if we think of leadership as guidance and responsibility rather than dominance. Having control of the resources brings with it a duty of care. Our dogs rely on us for food, shelter and safety. We have the knowledge and the wherewithal to take care of their health and wellbeing and to make sure they are well-behaved members of the canine and human community. In all these things, for sure, we need to take the lead.

But we can also allow ourselves to be the source of fun and play. It's just as important for our relationship with our dogs, but is not generally considered part of the 'pack leader' job description. So forget about the alpha stuff and just be a good friend.

MYTH 2: DOGS ARE PACK ANIMALS; THEY LEARN CONFIDENCE FROM OTHER DOGS

The other hangover from the 'dogs as wolves' mythology is that our dogs are pack animals who need to be with other dogs to thrive and that more timid souls can learn confidence from a more experienced, self-assured canine companion.

Dogs, like humans, are social animals and can get real pleasure from being with others of their own kind. Many dogs love to play with other dogs: you only have to see the crazy games of chasing and wrestling when sociable dogs encounter each other in the park or the enthusiasm with which dogs strain to get through the doors of doggie day care to recognise this. But not every dog loves all dogs or would prefer to be with them all the time. Some are highly selective about their playmates.

There is no real evidence to show that dogs are true pack animals anyway. Wild or semi-wild dogs don't form packs. They may form loose social groups for short periods of time but scavenge and live independently. Our dogs get together in the park, not because they're desperate to form a pack, but because they like to party.

Do dogs learn from other dogs? Puppies certainly learn life skills from their mother and their littermates. Most helpfully, from our point of view, they learn bite inhibition and civilised play behaviour from their siblings. If a puppy bites a littermate too hard or is too boisterous during play, the other pup will squeal and stop the game, so puppies learn that if they want to be able to play, they have to play nicely. And they continue to learn through experiences, some of which will be interactions with other dogs. This is why it's really important to ensure any interactions a young puppy has with older dogs are positive ones. Being snarled at by an older dog isn't the way to teach a pup good manners. That's another myth.

But can a fearful dog 'learn confidence' from another dog? This is unlikely, any more than someone with a fear of flying will overcome it just by sitting alongside a frequent flyer.

Again, scientific research suggests that in unfamiliar environments or worrying situations, dogs take more comfort from the proximity of a familiar, human caregiver than another dog and that the presence of their 'attachment figure' allows them to investigate the new more freely. An anxious dog might feel calmer with a canine companion, when they get to know each other, but it isn't guaranteed. They won't 'learn' to be less fearful from the other dog and in the absence of their canine chum, anxieties can return unless they have been addressed through careful training. And if a dog has separation anxiety, the presence of another dog may well not help at all – it's the absence of their human that causes the anxiety. In fact, adding another dog to the household can be just another challenging change in the environment for them to contend with. Helping them to adjust in small incremental stages to absences is the best way to progress.

MYTH 3: TRAINING WITH TREATS IS BRIBERY

This is another really unhelpful myth – that using food rewards to train is 'spoiling' or 'bribing' the dog. It's often quite tough for even the most loving, committed dog owners to put the idea of 'spoiling' entirely out of their heads. There's almost an inbuilt switch in the brain that says: 'Ooh, that's enough now.' Perhaps we get catapulted back to a childhood – a 'Don't eat all those sweets or you won't eat your dinner' state of mind.

But training obedience behaviours using treats isn't bribery, it's pay. You're rewarding performance and marking a job well done. In the initial stages of training a new behaviour, you need to be delivering the rewards fast and frequently. They don't need to be big so there's no worry that you're setting your dog on the path to obesity. As the dog learns the behaviour and begins to perform reliably on cue, then you can start phasing the treats down – that's down, not out. The occasional random payday is still important to act as an incentive.

BUSTING THE MYTHS

MYTH 4: MY DOG MISBEHAVES JUST TO SPITE ME

If the 'alpha' pack leader myth is loosely based on viewing dogs as wolves, the 'spite' myth stems from seeing them as all too imperfectly human. But this one has no more basis in fact. Sometimes people get frustrated that their dog chews their shoes or pees on the rug to 'spite' or 'punish' them because they've gone out. But dogs have no concept of spite – they simply do what works for them. If you're out and they're bored and there's something available that smells of you and works as a relaxing and entertaining chew, they're going to get stuck in. They don't see it as a favourite shoe – they have no fashion sense. They're not that big on interior design either, so those scatter cushions might be fair game as well.

It's not spite, it's opportunism, pure and simple. You can plan for this by giving them things to do and things to chew that you're happy for them to have – and by making sure your shoe collection is securely shut away. But for a small number of dogs, destructive behaviour might be real distress. If a dog shows signs of extreme anxiety every time the owner leaves, even for a very short time, and is regularly destroying furniture or woodwork, particularly around doorways, or soiling inside when left alone, this may be separation anxiety so could be time to call in specialist help.

MYTH 5: TUG AND FETCH ARE THE WRONG KINDS OF PLAY

It's tragic that more and more 'warnings' to owners about different kinds of play are appearing on social media. They're very effective as clickbait but have little validity beyond that. Two of the most common games that people are warned about are, of course, the very games that many play with their dogs and which their dogs love: tug and ball play. Play is so important to a good relationship with our dogs and isn't something we should start to get paranoid about. So let's kick these myths right out of the ballpark.

First, let's look at tug. This, so the myth goes, encourages aggression and if the dog is allowed to win gives them dominance over you (yes, we're back to that again). But tug is a great game to play if your dog enjoys it – it's good mental stimulation and excellent exercise. Letting them 'win' sometimes actually helps keep up the enjoyment and strengthens the bond between you – and it's only fair after all. Would you keep playing a game if you were always set up to lose? Tug can also be a good prompt for training, too. Teaching a 'drop it' means you can always wind down the game when you need to. (Though you will have to teach your 'drop it' first in a less exciting, high-stakes environment than a full-on game of tug.)

Then there's fetch. Some dogs can't get enough of this. Just the act of picking up a ball can send them into a tail-wagging, spinning frenzy of excitement. Why would we want to deny them so much pleasure? But, so the dire warnings go, playing ball can lead to Canine Compulsive Disorder or cause injury. There is no veterinary evidence that playing ball causes CCD and plenty that it is either genetic or stress-related. Sure, if you've got a real ball lover, your dog may want always and endlessly to play, but there's no need to make a drama or syndrome of it. If you can't or don't want to play, just put the ball away. They'll pester you for a while but will give up if you give them a distraction or just wait it out. I know this from very personal experience. Two of our dogs have been firmly of the view that there is no human activity that can't be usefully adapted to include a game of ball. And it is a valid point – working on the laptop, reading, TV watching, even abs workouts can all involve a game, if there's a will. But when we've had enough, we call time and they always settle. Boundless enthusiasm isn't clinical compulsion.

There are also warnings that ball-playing can cause injury from the repetitive chasing and the jumping and twisting actions if they try to catch a ball in mid-flight. This is possible and injuries do happen, sadly, but that's about being mindful of the environment and your throwing action. It doesn't have to mean a total ban. We can be sensibly risk-aware without being risk-averse and denying our dogs the fun of a game.

Of course there's a need for common sense: taking it easy in hot weather; matching your tug-pulling strength to that of your dog (do keep those competitive instincts in check); using the right size ball for your dog so they don't swallow it accidentally; and being more gentle with the ball throws for older dogs. One of the sweetest rituals I have seen was a man who brought his very elderly dog to the beach every morning for a game of ball. The dog would bark in excitement and the owner would roll the ball just a few feet across the sand. The dog would potter after it and bring it back, perhaps reliving in her head her heydays of premier league ball-chasing. Age doesn't need to stop play. But that's all there is to it – care and common sense.

MYTH 6: DON'T COMFORT YOUR DOG IF THEY'RE SCARED

This is another myth that really saddens me. It's an old one, but not a gold one and is pretty widespread. This maintains that comforting dogs if they're fearful or have had a stressful experience simply confirms to the dog that there is a reason to be fearful. I've never seen anything offered to back this up and it goes counter to all the evidence there is around fear in dogs – and humans, come to that.

Fear is an emotion. You can't reinforce it; it's not a conscious decision. Ignoring our dogs' fear does not make it any less real for them. If something has frightened them, it's down to us to get them away from a worrying situation as soon as we can and to provide reassurance. Ignoring the fear or repeatedly exposing them to an experience they find scary in a move to 'throw them in at the deep end and they'll get used to it' is more likely to exacerbate their stress.

Dogs take comfort from the reassurance of their humans. So give it: gently, quietly and calmly. Talk to them. Let them be close to you and give them a stroke, if they seek it, but don't make them feel trapped with big hugs if they're really frightened. The main message is: you're their friend, give them comfort.

If there's a particular situation or environment that causes them stress, you can help them gradually to overcome it through desensitisation and counterconditioning. But we'll talk more about that in Chapter 10.

MYTH 7: YOU CAN USE QUICK FIXES

We're in an age of instant solutions so it's perhaps inevitable that sometimes people think there must be shortcuts and hacks to training our dogs. Sadly, it isn't really true. Dogs don't generalise well, which means they can be quick at acquiring a new skill in the early stages of training, but will only apply that piece of learning very specifically – to a particular location, person, time of day, context. You need to practise in plenty of different situations until you've got it rock solid. If that sounds daunting, the good news is that once you've trained a few behaviours in this way, they'll start to generalise every new behaviour you teach without having to repeat the training in different contexts. If this all feels too tough, then cut yourself some slack. Your dog is a living, breathing, quirky individual. You don't need them to be 'practically perfect in every way'. You can focus on the big things that will make your life together safer, easier and more fun and manage the rest. When you've got those big things trained really well, then move on to something else.

There are no quick fixes to more serious behavioural issues, either. While a few can be swiftly solved with the right positive approach, most take time and patience. TV programmes (or YouTube channels) that give the impression that problems can be fixed in the space of an hour-long programme (including ad breaks) set us up for disappointment because it's simply not like that. You need to work at it; this is a long-term relationship, after all. But it's so worth it.

MYTH 8: THERE ARE NO BAD DOGS, JUST BAD OWNERS

More a mantra than a myth, this has gained currency in recent years and it's *partly* right. There are no 'bad dogs' in the moral sense; it's the 'just bad owners' part of this that is troubling. While there are some bad owners, no doubt, the idea that all behavioural problems are the result of bad owners is totally wrong. There are many reasons: genetics, early socialisation, experiences and, yes, occasionally, missteps in training. But let's not shame, blame or guilt-trip every owner who has a dog with behaviour problems. Many are trying to do their best for a dog that may not have had the best start in life. That deserves applause, not judgement.

CHAPTER 7

THE ACCIDENTAL TRAINER

Most dog lovers aren't born to train. While some really enjoy the whole meticulous process and get stuck into teaching tricks or flyball or agility with relish, when it comes down to it, the vast majority of us will settle for a relaxed, sociable dog we can chill out or cuddle up with and take for relaxing walks.

But however large or small our ambitions, when we get a dog, the job of training is thrust upon us out of necessity. Good, sociable behaviour doesn't come as standard and it can't be totally outsourced. So for all the accidental trainers out there in the world, it's all about making the most of the everyday opportunities to get some training started and reinforced.

EARLY STARTS

If you get a puppy, then socialisation and introductions to new experiences is the name of the game. Between three and sixteen weeks is a crucial time for development.[1] Some of the work should have been done by the breeder, but once you've got the pup settled into your home, it's down to you to be their guide in navigating the new.

Gentle introductions to as wide a variety of experiences as possible in a managed way helps build confidence about encountering the new later in life. This includes people in all their diversity: men (clean-shaven and bearded), women and those with different ethnicities, dress styles, heights and ages. Get them acquainted with other dogs, but make sure they're calm and friendly at this stage – being 'told off' by an older dog isn't part of the process. Introduce different sensations (sounds, smells, touch) and environments (rural, suburban, busy city centres, indoor and outdoor). You can shamelessly exploit the time of your family, friends, work colleagues, neighbours and random passers-by to help, but you'll need to get out and about as well. If a puppy is exposed to plenty of new things in their early months without incident, they become 'padded' – able to withstand the unexpected later in life. Then, even if they do encounter something that alarms, or even frightens them, they bounce back quickly.

The emphasis throughout your guided tour of the brave new world should always be on keeping the experiences positive and not overwhelming them with too much at once. No pushy parenting, please. Puppies need to take it gently. It's a similar principle to desensitisation with fearful dogs – except the stimulus in this case isn't a specific fear trigger, such as other dogs or men with beards, it's everything and anything they haven't encountered before. We want all these encounters to be rewarding.

Should you go for the formal and organised environment of classes? Puppy parties organised by vets in a sterile environment before they are fully vaccinated (which is when they can safely be allowed out to walk and mingle completely with other dogs) can help give a head start and introduce them to puppies other than their siblings and familiar humans.

Beyond that, puppy classes are, sadly, a mixed bag. There are some brilliant ones that are enjoyable and insightful, teach great skills and give you and your pup the chance to make new friends that could last a lifetime. But because there's no regulation in the industry, there are some truly awful ones too: classes that are an unholy mess of all the myths described in the previous chapter or where pups are manhandled and owners doom-schooled about the poor prospects for their learner hound without the firm hand of discipline. This is tragic. Pups are meant to have fun alongside their training – they're only canine kids, after all. Nevertheless, good puppy classes are THE best preventative measure you can take against behavioural problems later on because of the socialisation they provide. It's worth putting the effort into finding one. If you're really struggling, ask your vet if you can put a shout-out on their noticeboard and organise some socials for vaccinated puppies in the neighbourhood so that you can meet up for a walk and a play. There will be plenty of people in a similar boat. There are also plenty of entertaining and readable books by highly credentialled behaviourists and online courses in basic obedience.

Whatever route you take, doing the work of socialisation still means putting some structure into your outings to include plenty of different types of people and environments so that you widen your puppy's horizons as much as possible in the all-important early weeks and months. It's the most important gift you can give to your dog.

SECOND STARTS

What about older rescue dogs? The formative development weeks may be well behind them, but there could still be a training job of work to do, depending on what they've already experienced and whether they've been raised in a home or a kennel, a barn or a shelter or have lived on the streets. There's an often-repeated mantra about the settling-in period for rescue dogs: the 'rule of three' (three days to decompress, three weeks to learn new routines, three months to feel at home). This is wholly unrealistic. It makes no logical sense, given individual rescue dogs will have had very different formative experiences, and sets adopters up for disappointment and self-recrimination.

A dog that has had a loving first home or been in experienced foster care may settle more easily than a dog that has been used for breeding in a puppy farm or has been living on their wits as a stray. The range of stimuli to which they've been exposed and how 'padded' they are to the new is hugely important. Those who've had very limited human contact in their early months and exposure to very little stimulus may need a long, steady, patient approach, which can be many months, not days. They might need some considerable time to bond with their new family before being introduced gradually to new experiences, making sure each new encounter is positive, just as you would with a puppy. There is no neat formula.

Every dog is different. Some get their paws under the table and the household organised to their liking within days. Others need much longer.

You'll also want to take opportunities to get your puppy (or in some cases, your rescue dog) used to body handling, too. Start with gentle touches around neck and belly, ears, lips and paws with plenty of gentle praise and nibbles of treats. If your dog is particularly sensitive about some of this, don't force the issue. Start your handling somewhere else on the body and they'll get accustomed over time. All this will make it easier to do all the things responsible guardians need to do for their chums such as grooming, health checks, vet visits and teeth cleaning.

THE CARE NECESSITIES

When you've got past the initial orientation, and recognising that it may take a rescue newcomer some time before they're ready for the next stage, there is still more to be done. What are the big six obedience behaviours the accidental trainer absolutely wants to teach as standard? Clearly, it has to be those that keep you relaxed and in control and both of you safe in most of the situations you'll encounter day to day. My money would be on a snappy sit and/or down, a rock-solid stay, a loose lead walk, a leave it, a drop it and, above all, a rocket recall. A 'touch' cue, where you get your dog to nudge your hand on cue can be useful, too.

Because dogs don't generalise well initially, you'll need to apply the 'everything everywhere' principle, teaching in quiet environments first, but then starting the training again from the ground up when you move on to new and busier places. Don't be discouraged if all doesn't go to plan sometimes. You're asking your dog to do advanced-level, cross-species language learning. If what you thought was already trained has mysteriously broken down, you may need to repeat an earlier stage of the training to make sure it's really sunk in before taking the next step up. Just cut your dog (and yourself) some slack, take a step back in the process and try again.

For the accidental trainer, the best advice I can offer is to give yourself the headspace and a little extra time so that you can fit training into your life, be consistent and avoid mixed signals and frustration. And yes, I do know, that's easy to say, but less easy to do.

You can break most obedience behaviours down into bite-sized chunks – training snacks, if you like – that you can fit into daily routines and do in ten minutes, then come back to later. For example, if you want to teach your dog a formal 'look at me' walk to heel, you can practise for ten or fifteen minutes, come home, then go out again later. Getting a good recall can start with calling your dog from just a few feet away when they're already heading towards you, then for meals when they're already in the room and progressing to calling from elsewhere in the house, then from the garden. 'Training snacks' work well over time because your dog will process what they've learned overnight and will likely progress faster the next time.

Consistency is key to your dog learning your language and intentions, so don't muddy the waters if you can help it. Make training snacks as much part of the daily routine as taking a shower or brushing your teeth and make sure the whole household is on board. Set small incremental goals, rather than expecting great performance leaps. Let's say your dog gets so excited about the morning walk that they haul you out of the door, practically dislocating your shoulder in the process and regularly colliding with whoever is passing by on the street. You are going to need to apply a sanction every time until they get the idea that calm is the key to opening the door and getting out into the world. The first pull means moving back from the door. You can ask for a sit, if you've got that already very well trained. You'll probably only need to do this three or four times until you get a calm wait or a default sit while the door is being opened. You may then get an excited pull when the door is actually fully opened, but you can work on that next time. The objective of your first training snack is achieved. Do a 'waiting for calm' snack every morning at each stage of getting out of the door and the house until your departures are a thing of beauty and the morning walk happily anticipated by both of you. It may mean you only have time for a slightly shorter walk for a while, but that's not the end of the world. Don't fall into the trap of making training a battle of wills. Your dog isn't consciously defying you; they're just testing out what works best and most profitably for them. Persevere, be consistent and they'll get the hang of the rules of trade. Don't take it personally.

KNOW YOUR GOURMET'S PREFERENCES

The secret weapon of the accidental trainer is finding a reward that's really motivating. We humans can be really miserly on this front and it's a false economy. Most (though not all) dogs are easily motivated by food rewards, so when you're getting the basics mastered or grappling with a tricky behaviour issue, you want to find what they really love – no half measures. Ultimately, they all have their price. Food is a life necessity, after all.

Don't rely just on bits of the kibble they get as part of their regular meals. They get that without having to work their brain too hard, so think about training rewards as an extra special bonus. If they're enthusiastic about regular commercial treats, well and good. But to keep the performance graph on an upward curve, you may need to push the boat out. Find out what gets them extra attentive and excited. It will pay dividends in the end. It might be chicken or cheese, liver cake or sardines, dried sprats or frankfurter pieces. If the response to training is less than enthusiastic, up the ante. One of my clients was aghast that I was using premium Manchego cheese from the posh, local supermarket. Wasn't I going to spoil her, and bust the household budget at the same time? But this dog was a very, very picky eater and it took a lot to get her interested. If that meant Manchego, that's what I was going to use. You only need a little taste at a time (most accidental trainers use treats that are way too big). And it isn't forever. Once you've got your behaviours installed, you can, and should, start phasing the treats down, although remember that you'll still need to bring out a random bonus occasionally or you'll lose the behaviour. We're still talking about a trade agreement. Gourmet cheese won't be within the means of everyone – I completely recognise that. But be imaginative. Generally, the smellier the better is a good rule of thumb. You can always make your own baked liver treats for almost nothing. Your kitchen will not be the most popular hangout for your friends for a day after you've made a batch, but needs must.

But food is not the only motivator you can use to give training wings. There's also the power of play.

CHAPTER 8

THE POWER OF PLAY

One of the big 'truths' of science is that our understanding of it is never complete. Scientific law is certainly the best, most robust and repeatable explanation we have for something *as it stands today.* That's the point. For the most part, scientists welcome challenge. Science simply stands up and says to all challengers, 'if you can evidence a different way of looking at things, evidence that proves to be more robust than our current understanding, you can claim to have made a new advance'. Science isn't resistant to that, but you better make sure that you can stack it up. And that doesn't mean, as is sometimes the case in animal behaviour, dressing the established convention up as something new and innovative when the findings can easily be accounted for using the current knowledge.

The science of animal behaviour is rich and wholesome – sometimes densely packed and overwhelming. The tendency for the behaviour community to be complex is partly why I started writing this book, in the hope that we, as a community, can make the reality and principles of dog behaviour more accessible, interesting and enlightening to those who actually need to understand it. The basics of the science of animal behaviour should be the commonplace, the universally accepted, the place every dog parent turns to when they want to make their lives richer.

Those of us who get a kick from academic research can dig deep until we reach the Earth's core if we want to but let's not forget the importance of filtering that thick soup of knowledge to make it real and fit into people's lives. It's important to remember that people often have specific goals; they got their dog not as an academic exercise that has their brains imploding, but as the answer to a dream. Their dogs are not an interesting experiment or a course of study, they're a soulmate, a life companion, a family member.

If, as a community of evidence-based, force-free trainers and behaviourists, we can succeed in making that connection with our clients, loading them up with 'aha' moments, we can eradicate the myths and expel those who use pain and fear as a means of controlling dogs.

OK, I'll step down from my soapbox now, but I'm going to hold hard to this lofty ambition. The notion of being 'pack leader' and the need to 'dominate' your dog for fear of them usurping your powerbase has been around a long time and remains sticky, despite the attempts by behavioural science to debunk it – and debunk it we should. I believe the best route to that and to a world of happier coexistence with our dogs is through an understanding of how they learn. The big 'aha' moment is that however frustrating their behaviour might be, they are not doing it out of malice. They are just trying to speak to us.

The more I grow my knowledge about how dogs learn, the more convinced I am that we sometimes start our journeys with them in the wrong place, by focusing on the challenging behaviour. We need to take a step back, take a breath and focus not on the behaviour, but the relationship. If we do, we might just find that things go more smoothly as a result.

That is where play and affection come in. The power of play and the building of a bond has never felt more important to me than it does today. It's always tempting to go straight in for the behaviour change: have a specific plan with specific criteria steps to gradually work towards the behaviour we want. In the case of fearful dogs that most probably means starting a well-ordered plan of desensitisation (exposing them to whatever makes them fearful, but at a level that they find tolerable, which you gradually increase as they become accustomed) and counterconditioning (associating the scary thing with something positive so that it begins to predict good stuff and they develop a positive attitude to the thing that used to be scary). And there's nothing wrong with any of that. But when things begin to stall and we've checked ourselves for good technique and adherence to the plan and confirmed whether our client (the dog) is upset or not, perhaps the next thing to do is spend some time just taking notice of each other and exploiting what might be rare moments (at least at first) of connection.

There is no situation when this is more true than with those rescue dogs who are fearful of strangers. Their adopters, however well-meaning and desperate to make their newcomer feel welcome, are, for now, strangers themselves. That makes the whole thing very difficult. As far as the dog is concerned, the adopter is a new and scary person and the dog wants to maintain distance. But the adopter often hasn't got much choice in whether they get – as far as the dog is concerned – too close for comfort. They have to live in their own house. They have to cook, clean, sleep, prepare for work, accept deliveries and a whole gamut of things that will have the newbie rescue going 'Help! What in all that is terrifyingly incomprehensible is that?' All of which will probably push the new arrival wildly over their fear threshold, risking the dog becoming shut down or, at best, doing all it can to avoid contact.

So, you can see the problem. As the adopter, you're doing the feeding, providing comfortable sleeping spaces, opening doors in the vain hope that the trembling bundle of fur behind the sofa will go out and to the toilet. As far as you are concerned you are mum or dad or adopted brother or sister. But Greta Garbo behind the sofa doesn't see it that way. To them you are Godzilla or Kong or an invading alien army. Charming, right?

So what's to be done? Dropping treats to be collected at will is making some progress but is agonisingly slow. The dog, despite all your love and concern, seems to have little sense that you, the human, are part of their sanctuary, an ally who can be relied upon. They are sticking to the safe space that they have chosen. You can't intrude upon it except in urgent circumstances and, cowering there, the dog can simply deny your existence, metaphorically sticking their paws in their ears, closing their eyes and waiting for the terrifying thing to go away. These are the hard days and months.

The reality is that they are going to come out at some point. They are going to habituate to your presence and, as long as you don't move too suddenly, make a sudden noise, or there isn't a loud noise elsewhere, they may take a treat or two from close by or even from your hand, but we're a long way from introducing a lead or conjuring a sit on cue. Nevertheless, these are small breakthroughs. We shouldn't underestimate the beautiful significance of that first, tentative sniff of your hand from under the table or the gentle nudge of your knee because there's some small morsel left on your plate as you watch TV on the sofa. If you've ever felt that, you know just what a heart stopping, breath-stalling, magical moment it is. We have contact. We have communication across the species. You've bridged the chasm. In truth, for some that will signal an increasingly progressive march towards walks and snuggles on the sofa or fighting for the last sliver of the bed at bedtime. Others, however, will have the glorious sense of a journey begun, followed by the growing awareness that the journey's end is some distance away. If you are one of those people, give yourselves time to celebrate the win. It is a win – a huge win. You have been acquitted in the court of your dog's opinion of the charge of Ogre. You're not a friend yet, but you can turn around, thank your legal team and prepare for the press conference. You've done well.

But there is another tool in our kit to bridge the communication gulf and that tool is play.

Before we get too engrossed in just the sheer pleasure of playing with our dogs, it's worth digging into a bit of background. Why do animals play? How does play affect the way they relate to one another and us? Does the desire to play change over time? (And any other question I can think of or tangent I'd like to go on. I think you're used to my tendency to do that by now.)

> 'Play is foundational for bonding relationships and fostering tolerance. It's where we learn to trust and where we learn about the rules of gameplay; it increases creativity and resilience, and it's all about the generation of diversity of interactions, diversity of behaviours, diversity of connections.'
> – Isabel Behncke, primatologist and ethologist[1]

This quote by Isabel Behncke sums up the purpose of play in a wonderfully succinct way. We can, I think, take it as read that play has an intrinsic value of its own. That of just pure joy and pleasure. The bigger picture may involve learning about the world and how to cope with it, but that is achieved because the desire for play as fun is installed at the genetic level. That's what makes us compelled to turn to the earliest of mechanisms for learning. 'No motivation, no behaviour change', as Jean Donaldson reminds us.

For the most part, of course, puppies (like children) are blissfully unaware that their thirst for fun has a bigger purpose. But that's the artfulness of our learning and development 'engines'. It makes sense. We would soon lose interest and start to feel run down or even depressed if the entirety of the learning process was a chore. It's also why the nature–nurture debate is something of a fallacy. It's in our nature (genetically grounded) to be nurtured or to nurture. The ability to learn (nurture) is built in (by nature). Not everything in terms of building the toolkit for life is pre-installed, of course. We learn some of it on the way, but the ability to do that is a fundamental part of the genetic operating system.

You could sum it all up by saying that play is the process by which we acquire life skills, but I would hardly be fulfilling my duties here if I left it at that. Let's dissect Isabel Behncke's statement then:

'Play is foundational for bonding relationships ...' It's not by accident that this is front and centre. There is nothing more fundamental to a successful life in which an animal feels safe and able to navigate the complexities and uncertainties of existence than good relationships. Those relationships are both long-term and familial and

short-term and transient. In some cases, a dog struggles to form those healthy relationships and resorts instead to default fight, flight or freeze mechanisms to keep them safe in encounters with others. In these cases, it's likely they are haemorrhaging quality of life, and we need to intervene. Introducing and 'learning' play can be a strategy to do that.

'... *and fostering tolerance*' My interpretation of that would be: putting up with stuff. Holding it together in order to preserve the status quo, even if it's a little TOO much. For 'tolerance' I read the ability to absorb less than welcome, but essentially benign encounters or experiences.

'*It's where we learn to trust and where we learn about the rules of gameplay*' In essence the premise of this book is that trust (or trusted familiarity) is the bedrock of building confidence and resilience in dogs. Playing with each other and with us is the elixir of a stable, happy and well-padded life.

'*It increases creativity and resilience, and it's all about the generation of diversity of interactions, diversity of behaviors, diversity of connections*' Play gives us the ability to experiment and explore the world around us. It enables us to absorb a knock or two and bounce back and it helps us take the new and unfamiliar in our stride. It encourages the broadening of the 'friend' pool, the widening of the social network and hones the behavioural vocabulary that allows that to happen with minimal risk. It builds the skills that allow peaceful engagement with others, whether that coming together is lifelong or transient.

THE POWER OF PLAY

Play does an enormous amount of heavy lifting in the life skills development of most species and yet, as humans, we tend to dismiss it as trivial, purposeless, something to put aside with childish things so that we can get on with the important grown-up stuff. That feels like a mistake, but, happily, our dogs can help reconnect us with this important skill. Because it's worth emphasising that play is a skill to be acquired and practised to make perfect. The inclination to engage with it may be pre-installed but becoming an aficionado takes practice, practice, practice.

We can learn a lot from observing dogs playing together, assuming they are sociable and want to do so. Dog play has a language. It is a way for dogs to communicate their intentions to one another and ensure that play is a pleasant experience for all concerned. As an aside, adult dogs who don't play or are selective about their playmates do not necessarily have a problem with other dogs. Their inclination to play may just have reduced over time as they have grown older and the robust rough and tumble of a play party doesn't suit them anymore. Of course, they might have an issue with unfamiliar dogs in which case, in conjunction with other indicators that a qualified professional can help you with, the absence of play can be very telling. But for now, let's stick with dogs who do play with others and break down what we are seeing. The first thing to say is that if you have witnessed play that involves fighting, biting, growling and barking it isn't time to panic and withdraw your dog from the local play group or day care.

You just need to know how to read the messages and test that you are right about them.

As already mentioned, play is a rehearsal for the life skills that any animal will need for a successful future. Because of that, you are likely to see a range of behaviours during dog play that look a lot like aggression. Fighting, biting and growling are all commonplace in healthy dog play.

Predatory behaviour like chasing and pinning won't be unusual either. Predation, for the record, is sometimes bundled in with aggressive behaviours, but I would argue that the crucial elements of aggression are missing. It looks violent for sure – if you've watched wildlife programmes on the television, then you will be in no doubt of that. But consider this: predation is food acquisition, plain and simple. It's the animal equivalent of going to the supermarket and the fact that violence is present in the act of food acquisition doesn't mean that your dog is going to attack the neighbour or the kids. It's just picking up some Cheerios from the cereal shelf. It makes no difference whatsoever that you are already supplying food in plentiful quantities. The behaviours involved in hunting will still be rehearsed in play and I would be more concerned about their absence than their presence.

Finally, you are going to see behaviours that are associated with reproduction such as mounting.

I'M JUST HERE FOR FUN, YOU DIG?

So, what to look for? My friend and colleague Jane Sigsworth describes it by way of a nice pithy acronym: MARS. This, Jane explains, stands for meta-signals, activity shifts, role-reversal and self-handicapping.

Meta-signals are behaviours designed to indicate that the intention is play. They can come before play by way of an invitation or during play to indicate that all is well and the intention continues to be play. Typically meta-signals will include play bows, bouncy exaggerated movements, play faces and paw raises. Let's unpack some of these. Play bows, as we saw in our chapter on dog body language, are when one or both dogs drop their front ends down close to the floor by spreading their front legs wide. Their rumps are left in an upright position, thus presenting something akin to a bow.

Activity shifts are a sign of healthy social play. You should see the engaged dogs changing the activity, from fighting to chasing, for example, and these changes are generally punctuated with meta-signals so that everyone knows that the change is still play.

Role-reversals can be a little trickier. They can serve as a good indicator of healthy play: 'I'll be on the top of the fight, then let's swap and you can be on top' or 'I'll chase you, then you chase me.' It may not be consistent, however. Some dogs actively prefer certain roles and their playmate may well be happy to oblige. This doesn't mean that things are going wrong provided meta-signals are present. Pauses in the activity are a clue that the play is OK too.

It can be difficult, if you are new to watching and supervising dog play, to know when things are going well and when it might be getting a little out of hand. But there is a 'safety valve' you can employ to help you. If in doubt, use a consent test. This tells us how comfortable a dog is with the style of play that they are involved in. Simply take the more robust player out of the rumble and see what the 'victim' does. Do they look for another playmate, suggesting they would like a less crazy time of things, or do they choose to go back to play with their favourite goofball? If it's the latter, then all is well and you can let things continue. There's no downside to a consent test so you can do them as regularly as you feel the need.

Another thing you can look out for, particularly if there is a mismatch either in size or play style, is self-handicapping. This is a really interesting self-intervention by the more vigorous or robust player towards a smaller or more timid character. It's also the sign of a skilled and experienced dog. Self-handicapping is when the bigger or more robust dog moderates the energy in their play to match that of the smaller or more gentle dog. It often results in exaggerated, quite ritualised behaviours that are wonderful to watch, so keep an eye out for it when you are watching mismatched play.

CHAPTER 9

WHEN GOOD INTENTIONS GO BAD

Most of us start with the very best of intentions when we get a dog – perhaps the very largest of ambitions too. We fantasise that we're going to have a super-smart sidekick who'll come at the first syllable of their name, trot obediently at our side, know all their toys by name and do tricks to amaze our friends, rather than planting muddy pawprints on their jeans. This is, of course, all possible, but there's an awful lot of diligent effort needed to achieve it. Many of us, however, don't need elite canine performance. When we come up close to the reality of our dog's quirks and personality and our own time and attention span, most of us are happy to settle for good. And that's just fine.

But sometimes even that limited goal seems to be endlessly subverted. This is all because of the way dogs learn. They do what works best for them. It isn't malice, it's how they've evolutionarily developed to thrive and survive.

Given that we are (allegedly) the superior species, how do they manage to outmanoeuvre and outwit us?

THE DOG AS PREDICTIVE GENIUS

The first reason is their powers of prediction. They are almost uncanny soothsayers. Their super-sharp senses help them detect changes in their environment that we hardly notice. They can sense a turn in the weather, even before we're conscious of the pressure dropping, the wind rising or the first drop of rain falling. They will notice subtle differences in a familiar environment that we barely register.

They're keen students of our behaviour, too. Far more than most humans, who are endlessly distracted by trivia, dogs pay very close attention. They have a dedicated area of their brain devoted to processing human faces.[1] They learn our 'tells' with a precision that would be the envy of a professional poker player. They detect shifts in our mood through subtle signals in our body odour. They are acutely tuned in to the small signals of intention we subconsciously give out.[2] Their senses and skills in observation are what make trained assistance dogs so good at alerting their human companions to the onset of a seizure or when blood sugar is getting dangerously low.

Dogs are also clever at tracking whole complex sequences of events so that they can 'predict' what comes next. Our dog, Ripley, is well attuned to aspects of our daily routine. Each morning, the barely audible click of the mobile phone on the bedside table being unplugged from the charger provokes an excited explosion of wagging, wiggling and snortling, because she knows this means 'ball on the beach'. At the point of that click, no one has actually got out of bed, let alone got dressed to leave the house. There are dozens of individual actions separating that sound from the first throw of the ball, but she's tracked back. It may be a journey of a thousand steps, but ball on the beach starts with that click and if she hustles things along, it will come all the sooner.

Dogs' ability to learn complex chains of behaviour can be very helpful. It's how enthusiasts teach their dogs to run complex agility courses and how service dogs are trained to help their people with household tasks. They learn not just that 'A' predicts 'B', but that it's followed by 'C', 'D', 'E' and on through a long, extended sequence.

This is how things can go so awry for us unwary humans. If a dog learns that the end of a sequence is something they don't find very desirable, they'll try and avoid it. For example, if your dog is a lover of wallowing in muddy puddles (or worse), but less of a fan of the bath that results when they get home, they may make a connection along the chain of events. And the connection probably won't be the one you'd hope for. It won't be: 'If I stop rolling in mud, I won't get a bath.' Oh, no. But it might be: 'When I get back into the car after a great walk with a really satisfying roll in something gorgeously smelly, I get taken home and hosed down. I hate that. I know – I just won't get back in the car.' That, of course, now leaves you with a whole new olfactory challenge to overcome.

Equally, you can find yourself deep in the land of unintended and unwanted consequences when the end of the chain is perfectly desirable from your dog's point of view, but one of the steps along the way is less desirable from yours. Let's say, for example, your dog always goes demented, barking at the window, when anyone passes by on the street. It's loud, it's nerve-jangling and it's deeply unpopular with the neighbours. So you set out to train 'quiet'. You are diligent. Dog sees passer-by, dog barks, you offer a treat, say 'quiet' and when barking stops instantly give a treat. After a while, you can stop the barking on cue. But unless you get in first with your 'quiet' and do it regularly, your dog will still *start* barking every time they see someone going past the window because, to them, that's part of the chain that leads to the reward.

More seriously, the law of unintended consequences in the way dogs perceive the individual links in a chain of events makes aversive training methods potentially dangerous, as well as cruel. If a dog is given a jolt from a shock collar for an unwanted behaviour, the connection they make may not be with that behaviour. Instead, they might make a link to something else that caught their attention at the exact moment they felt the shock: the car passing, the children laughing on their way home from school, the woman walking the Bichon. A negative association is created between the shock and any one of these unrelated events, so the dog begins to fear cars or children laughing or Bichons. I say 'fear' advisedly, by the way. Shock collars work to change behaviours because they are an unpleasant experience – not just a little vibration to get attention. That's the marketing. They deter dogs from behaviours because they scare or, potentially, actually cause pain. And if a dog makes a random connection between that sensation and something else going on in the world at the time he feels it, rather than linking it to the 'unwanted' behaviour, you are creating a world of problems.

TIMING IS EVERYTHING

If we want to harness our dog's observational and predictive superpowers to teach them the kind of behaviour we want, timing is everything. Our dogs don't come with our language pre-installed on their hard drive, so they have to make the link between our words, their actions and the consequences. On the whole, they try quite hard to do this. Not because they're simply 'anxious to please' – even when they're putting on that cute expression – but because they want to do something that's profitable for them. Initially, there's a lot of guesswork and trial and error on their part. We can lure them into performing an action, but their only real tip-off that a particular word means a particular behaviour is when we show them that they've done the right thing the very instant they do it – with a word of praise and something they really love. Unless we get the timing right consistently, they just end up confused. While it can be entertaining to see them go through their repertoire of all the things they've learned that, based on past experience, gets you to stump up a reward, it doesn't get you much further forward. Bad timing is the graveyard of good intentions.

THE INFLATIONARY SPIRAL

Then there's the inflationary spiral. Back in Chapter 3, I explained how training and behaviour are, basically, economics. And within any dog-human household economy, there lurks the possibility of rampant inflation.

It generally starts getting out of hand because we're in a hurry. We want the dog to get in the car or abandon that sniff and get on with the walk or come in from the garden because we want to go to bed. And we want this to happen NOW. In our need to move things along, we put our hands in our pockets for the tasty treats. We pay in advance for the behaviour we want and, if we're really under pressure, we'll probably pay over the odds because it seems the easiest thing to do at the time. We get to crack on with what we need to do and the dog gets an extra treat or two. Win-win, right? Unfortunately, smart dogs see the potential here for a life of unlimited bonuses, entirely unrelated to performance. When you've given in once, those savvy canines are just going to keep coming back for more.

Using food rewards as a lure when you're doing initial training is perfectly valid, but in these circumstances, it's storing up a whole barrel of trouble.

WHEN GOOD INTENTIONS GO BAD

So let's just imagine you've let your dog out in the garden before bedtime and they've dawdled at coming back in because they've found an interesting smell left by the neighbour's cat. You've had a long day. You're tired and fed up. It's cold and you don't want to leave the door open, waiting for them to come back in. But needs must, so you go down the garden and start scattering a few treats as a trail to lead them to the back door. They happily hoover them up and come in. This, as far as your dog is concerned, is a new and welcome development in the bedtime routine. The next night, they might just test the water to see what happens. They'll hang out for longer, maybe skitter off to the end of the garden in the dark. You try and wait it out for a while and then give in. Before you know where you are, it's become a nightly ritual and the price of compliance with your pleadings to come in from the garden has gone dramatically up.

Inflationary spirals start when we forget that positive reinforcement is just that: reinforcement. It's rewarding a behaviour performed, a job well done, not offering an advance against future compliance. I make no lofty judgement about anyone who's been caught out in this way. We've all been there at one time or another. (Our rescue Rottweiler, Murphy, was a master at this particular form of manipulation.) But rolling back the inflationary spiral takes a painful amount of patience and it's a challenge best avoided where possible. Try and keep front of mind that this special relationship, loving as it is, is based on trade, not aid.

While it's important not to fall into the trap of advance payments, it's perfectly acceptable to run a tab with your dog and expect plenty of good behaviour 'on account'. If a dog isn't fearful and has really, fully learned a behaviour, you don't need to pay every single time you ask for it. You can just put it on the tab. In fact, randomising rewards can incentivise your dog to keep focused because this might be the occasion they'll get a big payout. Rewarding every time can take the edge off performance but you can't keep endlessly running up credit. You do have to pay your tab, now and again, otherwise your dog will, perfectly reasonably, withdraw service.

THE STICKY STUFF

While our dogs do appear sometimes to be soothsayers in their ability to anticipate what's next in a familiar sequence of events, they aren't mind readers and they do see the world differently to us. Lots of the learning they do is very specific to context and environment, and we have to acknowledge and work with that. A 'sit' in the house is a different beast from a sit in a busy park or a lively pub or café where there's an awful lot going on. It might even be different if the 'sit' request is made by someone new in a different tone of voice.

Words that might convey the same meaning to us, clearly don't for our dogs. You say 'stay', your partner says 'wait'. Tomay-to, tomato? Not really. You might get lucky if the word sounds very similar, but you're asking a lot. Consistency, not creativity, is king. Same hand gestures, same words, same tone. It may sound obvious, but we're so used to shaking up our vocabulary in daily life, it's hard to get out of the habit.

Making the learning sticky also means adjusting the pace of teaching to the dog's pace of learning. They need to be performing the action in less challenging situations before you can move on to the next level of difficulty. Can they sit-stay while you take a couple of steps away and do that consistently time after time? Then take the level of difficulty up just a notch. Our good training intentions go bad most frequently because we don't have the patience to break down the stages of learning and do the 'reps'.

HARMONY RULES

Whatever the occasional frustrations that come with life with a dog, it's always worth taking a beat and thinking about what counts. It's a sadness to me that the dog–human relationship that should be about simple, happy harmony gets loaded with so much judgement. There's always someone tutting about what your dog should or should not be doing. 'What, your dog doesn't even know how to give paw?' 'What, you let your dog sleep on the bed?'

If your good intentions haven't got you as far as you'd hoped, put the 'shoulds' on hold for a while and think about what really matters. Consider progress in terms of what you can do today that you couldn't do last month and what that means for what you could do next week. Just as it's easy to slide into bad training habits, it's even easier to fail to see how much progress you're actually making.

Don't be discouraged by the setbacks. Remember your dog is driven by simple motivations. They're not trying to defy, cheat or gaslight you. They're just trying to live their evolutionary best life. Just like you.

MURPHY

Our boy, Murphy, was a world-class grifter and master manipulator. In the mutual trade agreement of coexistence with us, he was, without doubt, an expert negotiator.

We adopted him at eighteen months old. Like Thomson, he was a Rottweiler, but much less of a challenge. His people had rehomed him reluctantly when they had to move and couldn't take him with them. They'd taught him some lovely manners.

However, his Moriarty tendencies emerged pretty quickly. We'd had him just a couple of days and were making a full-on roast Sunday lunch. This was something we did very, very rarely and we were looking forward to it. The joint was on a chopping board on the kitchen counter, resting while we sorted out the gravy. This was going to be good.

As far as we were aware, Murphy was in the living room, but when we turned round to get the plates, there was Murphy behind us, standing proud with the meat clamped firmly in his jaws. His expression said clearly that he had made absolutely the right choice in coming to live with these humans. We contemplated the tray of roast potatoes, the vegetables and the pan of gravy and wondered, momentarily, if we felt the same. His nice manners, sadly, didn't include a 'drop it', which, under the circumstances, would have been pretty miraculous. Surprisingly, however, he was eventually prepared to trade. We didn't eat the recovered joint, of course; we reclaimed it largely to salvage the remnants of our dignity, not our dinner. We were only two days in and already he'd got our number.

We learned to be more careful, but he learned too, and he was a fast learner, always with his eye on the main chance. And the person he identified as most ripe for profitable manipulation was my partner, Nina.

I'm not training-shaming here. She'd be the first to admit that she was often in a hurry to get on – places to be, calls to make, trains to catch – and gradually slid into a pattern of luring Murphy when he dawdled and delayed. And Murphy was storing all this up in his opportunity database. Maybe his super-senses came into play here as well – he could almost certainly tell from the body language when she was getting tense because she was late for something or other.

But the first real upward curve of the inflationary spiral came when I was in hospital with appendicitis. Nina was working from home and, in a gap between Zoom meetings, decided to take Murphy out for a walk in the hills. She bundled him into the car and headed off – plenty of time for a decent ramble before she needed to get back. But at the end of the walk, when she called him to get into the car, he stood a few feet away from the boot and refused to budge. So she did what she'd done before and chucked a couple of treats into the crate to get him moving. But this time, it all started to go wrong. Murphy made no move to get into the car. Instead, he gave her the cool, appraising look of a con artist sizing up a potential mark, concluded she was good for an awful lot more than a couple of treats and sat down. Pleading, cajoling, coaxing, walking him away on the lead, then turning back and taking a run up at it, nothing worked. By now, panic was setting in. So she emptied the entire contents of the treat pouch into the back of the car. Murphy jumped in. Crisis averted.

Of course, this was not the end of the story. The next time, the price of getting back into the car at the end of the walk had gone up. A handful of treats wasn't going to cover it. She had to raid the emergency supply of cheese oatcakes from the glove compartment. Within days, inflation was out of control. She was leaving to go on walks with treat bag and pockets stuffed with anything and everything that might tempt him into compliance when the time came to head for home. At its peak, the cost of getting Murphy into the car after a walk was a handful of treats, a piece of Manchego cheese, two oatcakes and a quarter of a lamb and rice stick. (Personally, I think he was selling himself short – he should have held out for the whole lamb and rice stick.) The prospect of a crash in the household economy loomed (not to mention busting through his harness with all the extra calories).

Buoyed up with success, Murphy tried to overreach and exact payment for getting out of the car when they reached home as well. But by this time, Nina had reached her limit. She stood firm and waited him out. After that, while getting in the car at the end of the walk was still extortionately high, getting back out was thrown in as part of a two-for-one deal.

When I was back, I did manage to get reward inflation to drop back to below zero, but for Nina it was a hard lesson learned. Never, ever make advance payments to a clever Rottweiler. He'll fleece you for everything you've got.

Murphy stayed a hustler his entire life and I was not immune to being played occasionally, even though I should have known better. Keeping one step ahead of him meant always keeping your wits about you. But no matter how many times we pledged over candlelight and the memories of ancestors long dead that we would just ensure that the opportunities for Murphy to exploit us would be so infrequent that they'd be manageable, we were always going to be outclassed. In my defence I had only just started my education in animal behaviour at the time, but Murphy was entirely unwilling to make any allowances.

He learned very early on that the TV remote was a highly valuable item in the house. It likely had something to do with the frantic scrabbling and panic that followed when he casually lifted it from wherever one or other of us had left it. That was the tip-off. If something tasty wasn't immediately offered in trade, he'd gradually close his jaws a few millimetres. The gentle cracking sound generally increased the level of urgency.

Murphy loved everything about this game: the nonchalant approach, the theft itself, the chase around the coffee table (I suspect his favourite part), the ransom negotiation and the ultimate pay-off. No cops, no cameras, no talking.

As time went by, again the price for release invariably climbed. The inflationary spiral took another victim; Murphy understood the principle of supply and demand all too well. We tried extinction trials. We bought cheap, sacrificial remotes (batteries removed) and left them lying around to show just how little we cared about the twilight raids across the coffee table or sofa. It worked, or so we thought. The appeal of the remote began to decline as stealing them became unprofitable. And yet ... when the real thing came out and got carelessly, nay criminally, neglected, he knew. As we lapsed into smug complacency, thinking we'd got this sorted, Murphy was diligently watching and waiting.

One day as we were watching TV, with Murphy snoozing in his favourite chair, we found out just how much of the long game he was capable of playing. I had put the remote on the back of the sofa for ease of access for us, but not for Murphy. Nina had put her phone on the coffee table in front of us. Murphy had never shown any interest in phones before. Curious! But this time, he slowly stretched and unwound himself from the chair in a way that a dog does when they're just going to settle down in front of the fire. Only this time, Murphy was putting the long con into operation. He'd been waiting months for just the right conditions and now was that moment. As he sauntered past the table, he reached over and took Nina's phone off the low coffee table and just stood there. If dogs were capable of chuckling, that's what he would have been doing: 'Gotcha'.

Both Nina and I jumped up from the sofa and made for the kitchen. We weren't going to mess about. We were going to have to trade and we were going to have to go large. As we jumped up and made for the kitchen door, Murphy, in what seemed like one movement, dropped the phone (entirely unmarked), vaulted the coffee table on to the sofa and with an elegant flick of his head grabbed the TV remote. The price had just gone up ... again.

CHAPTER 10

THE FEARFUL DOG

Fear is a bit like a computer virus that gets into your machine simply by clicking on a link. Fear, in other words, is shockingly easy to install and hideously difficult to uninstall. It often leaves something behind and, unlike a computer virus, it's not always obvious what's causing the problem. And because our dogs are not robots the cause can be varied and the road to recovery pitted with the unexpected. That isn't to say that there isn't a solution but – and there's no getting away from it – dealing with fear in dogs is a tough gig!

There's one unhelpful myth around fear that I want to bust, largely because I would really like people who are living with a fearful dog to seek support and help as they travel through such an uncertain landscape. That is the idea that all fearful dogs have been mistreated or abused. While this can be true, it's far from true for all of those who are identified as fearful.

Fear may start life as the result of a single unfortunate event intended by no one. It may be a single unkind act or a prolonged mistreatment. But it can also be the product of genetic predisposition, often made even more likely by human selective breeding for traits that seem useful to us but are rooted in a heritage of suspicion or mistrust. It may be a genetic by-product – a single traumatic event experienced by the pregnant mother can have dramatic consequences for her litter, as can the behaviour of a nervous mother during pregnancy as well as during the formative early weeks of the litter's life.

Experiments on mice have shown that nervous mothers often give birth to nervous litters while confident mothers, in turn, give birth to confident broods. However, if you swap those litters shortly after birth the fearful dam will raise fearful young, while the anxious youngsters reared by the confident parent remain fearful. Easy to install, devilishly hard to get rid of. It is an absolute minefield and one where we've learned that many of the old adages don't apply. For example: 'Throw them in at the deep end and they'll learn to swim.' Sadly, they'll very likely sink. The language barrier between dogs and people is challenging enough without us misinterpreting expressions of fear or anxiety as 'bad behaviour' that needs to be punished out.

Our dogs can't pick up the phone and ask for help from a mental health professional. They can't tell you what's on their mind and, for our part, sadly, we can't just put a description of their behaviour into Google Translate and get back an appeal for support. All too often fearful behaviour, whether it includes aggressive behaviours or not, is misconstrued as belligerent, confrontational or competitive. The reality is that dogs have a limited 'vocabulary' with which to explain to us how they are feeling. Their vocabulary is made up of a set of behaviours that we as a species (the smartest on the planet) need to get better at understanding, all so that our friend, the dog, can access better services that will aid them towards recovery. The long and the short of it is that we are ultimately responsible for making dogs who they are today through domestication and collaboration in selective breeding. They are way more than 'just dogs'. They are colleagues, friends and family members. They have identities and personalities. They are sentient beings who deserve our respect, tenderness and understanding. That means learning how they learn and making sure that we put that knowledge to the service of not just making our lives easier with them, but doing everything in our power to make their lives with us more joyous every day.

SAFE SPACES, SAFE BASES AND THE TRUST CONUNDRUM

Many people will be familiar with the idea of safe spaces both for people and dogs. Those who have fostered or adopted rescue dogs are probably very aware of the need for these dogs to establish their own safe space when they join a household. Anyone whose dog suffers from sound sensitivity, such as fear of thunderstorms or fireworks, will probably know all about this too. It's crucial when a dog is anxious or outright afraid to let them find their safe space. The safe space must be reliable. i.e. available at all times. Nothing is gained by blocking off or dismantling it in order to encourage more engagement or exploration. The effect is most likely to be the opposite, namely a shutdown or a panicked dog and instead something intended to increase attachment to the home and the human results in an erosion of trust.

Let's talk about that word 'trust' for a moment. I confess that it's one of those words that I'm a little bit twitchy about when it comes to building relationships and attachments with dogs. Why? Because there's risk in it for people who are already dealing with a fundamentally difficult problem. The risk is that every setback or plateau makes them feel that their dog doesn't 'trust' them; that there's something they have failed to do in setting the relationship on the right track. That they are, in essence, 'untrustworthy'. What other explanation could there be for their dog (who they love) not trusting them? The answer to that question, in truth, is that there can be plenty of reasons for a fearful dog maintaining their caution and it isn't necessarily something that we, as the human caregiver, have consciously done or neglected to do. While it is true that the ultimate outcome from building attachment, safety and confidence can be characterised as 'trust', it isn't accurate to characterise the absence of that confidence as the same as mistrust. That puts the focus in the wrong place. It puts the onus on the character of the caregiver when it should be on the fear.

Fear is not always rational or logical. Sometimes it makes sense to be afraid of things that are genuinely dangerous. At other times the fear (and the behaviour it evokes) can seem nonsensical and even detrimental. We need to focus on the fear and how to reduce it and not take the regressions or the stalls as a personal affront to us. It simply isn't helpful for dog or human.

Here's an example. I described earlier how I introduced Thomson, my first Rottie, to my partner Nina. Thomson was scared of strangers; if one got too close, they were likely to learn pretty quickly that giving Thomson a wide berth was a good plan. Thomson had no history with Nina. He hadn't hired a private detective to check her out; there was no secret dossier that concluded she was not to be trusted. It wasn't personal. Nina was simply unfamiliar and, therefore, innately scary.

Why? I've no idea. There were elements of his history that might explain it. But the 'why' as we've discussed before, isn't important. The objective, regardless of the reason for the fear, was to make Nina less scary and we did this through a process of desensitisation (brief interactions at a distance and duration that he was comfortable with) and counterconditioning (associating Nina's presence with things he loved). As Nina became reliably predictive of good stuff like cheese or ball games, so she became less scary, more familiar and, sure, ultimately, 'trusted' – even loved. The pitfalls and setbacks that inevitably punctuate training protocols for fearful dogs are not about erosion of trust. They're not failures. They are the normal course of progress. They give us important information by indicating that more time at that step is needed, or that we've pushed a little too early and the current step is beyond our dog's ability to cope. After all, our job with fearful dogs is to help them learn that the world (and the world with us) is fundamentally safe. My message to anyone reading this book and who is currently dealing with a fearful dog and is looking for a force-free way of solving the problem, is: 'Respect, you're doing a great job!' And if you're finding it tough and need help, or even just a sounding board, then find a specialist to help.

BASES AND BUBBLES

The concept of a safe space is not a difficult one but it is, of course, limiting when it comes to gently expanding experiences and horizons. Yet with time and infinite patience it's possible to make a fearful dog's caregivers become a safe base, providing reliability and reassurance in navigating the new and making the safe space, essentially, mobile. It's important not to rush into this, however. It's a step-by-step process.

The choice of a safe space seems almost instinctive for a newly arrived rescue dog. It's often the first place they come across that they can shelter under or behind so they can defend themselves from danger if they need: under the bed, squeezed into an obscured corner of the room or behind a chair or sofa. As they become assured, over time, that the space does indeed provide safety, thanks to the absence of intrusion, it becomes the nucleus of their world. It's the place they can retreat to in the knowledge that it offers sanctuary and harbours no danger. As time goes on, they might cautiously explore the environment outside the safe space in widening circles, but always with the route back at the forefront of their mind. If the dog is allowed to do this unhindered, the wider environment may well begin to be perceived as safe (usually the interior space in the home) and the people in it, having presented as little risk as possible, might, by this stage, be viewed as neutral. Leaving the dog be, if that is all we do, may well mean the human presence remains neutral at best. It's something akin to habituation. That's OK for a while. But we can have a bigger ambition – to establish a more flexible and mobile safe and secure base, one associated with the human caregivers that allows them to introduce their hyper-vigilant dog to a wider circle of new experiences while maintaining their sense of safety. For that, we're going to need to employ a more active strategy.

Dogs rely on predictability to learn efficiently. Classical or Pavlovian conditioning depends on the solid reliability that one thing will predict another. For operant conditioning to be effective, the dog needs to be able to predict what the outcome of his actions will be, reliably and confidently. In this sense, predictability and safety are natural bedfellows.

Because the safe space is established as predictable, secure, unthreatening, reliable and accessible at all times, it acquires value. To acquire a similar value, a secure, human attachment figure needs to have more than just neutral status. This takes some thought and care. For a very fearful dog, humans only stay neutral in this uncertain world if their behaviour is predictable and uncomplicated, i.e. not moving, static – basically not doing anything new. Even something as simple as a movement the fearful dog hasn't encountered before, can make the human become unpredictable or, if you like, 'untrustworthy'. They haven't done anything consciously scary, they've just done something different.

In this fragile environment, how do we transfer that elusive elixir of 'security' from the safe space to the person? How do we make the static figure into a secure attachment base? The short answer is 'carefully'. It is perhaps an overused phrase when working with fearful animals, but the tortoise really does win this race. There are no shortcuts in bond development between dogs and people, any more than there is between one person and another. There are no quick fixes to fear conditions. Since that is the case, we might as well do everything in our power to add stability to the foundation. That, to a large extent, means thinking how we can exploit every moment we spend with our dogs in order to grow the attachment we have with them. What do they love and feel positive about? By doing so, we can steadily construct a bubble of confidence and 'trust' that our dogs can share with us as we navigate the world together. If we can do this, that original safe space becomes just another comfortable place, as the need for the security it offers reduces. And bear in mind that if your bubble ruptures at any time and leaks safety, it isn't personal. It's just that the fear the dog feels currently trumps the attachment – not because of you, but because of whatever at that moment evoked the fear. Fear doesn't go away simply because we tell it to and sometimes it becomes overwhelming, despite our best intentions. That is a signal to withdraw back to a point where the dog feels safety and proceed more slowly next time.

Our aim is for the base to always be stable, but it's important to acknowledge that it may not always be that way and be alert to the indicators that tell us we're sailing close to our dog's fear threshold.

MOVING THE THRESHOLD

'Thresholds are dangerous places, neither here nor there, and walking across one is like stepping off the edge of a cliff in the naive faith that you'll sprout wings half-way down.'

– Alix. E. Harrow, *The Ten Thousand Doors of January*[1]

In dog behaviour, a 'threshold' is a concept in classical conditioning; more specifically, it references when dogs are upset: that is either over or under a fear or anxiety threshold. It is not a term found in the practice of operant conditioning. I sometimes see it used in relation to dogs who are clearly not upset, but simply exhibiting excitable or (inadvertently reinforced) unwanted behaviour. This, as an aside, is unhelpful. Professionals in the field should understand the terminology well enough to use it consistently, both with other professionals and with clients.

The object of the exercise when working with a fearful dog is always to avoid exposing them to stressors that take them over their fear threshold, the state where they start to exhibit fearful behaviour. What might that look like? Well, the lines can be blurry, for sure. There are varying degrees, which means that the quotation above, while painting a useful picture, may not serve entirely to illustrate the point. If a dog is trembling, salivating and looking desperately for sanctuary, potentially exhibiting aggressive behaviour, then wings have clearly not sprouted and you'll need to employ some significant measures urgently to get them back to a sense of safety.

More problematically, serious distress may be seen in shutdown behaviour. The dog does nothing. They may even appear calm, but that is far from the truth. From the triumvirate of escape strategies (fight, flight, freeze) they've deployed freeze. It may even be the case that what you're seeing is a phenomenon known as 'learned helplessness'.

This is when the dog has been subjected to a frightening outcome so frequently, with no prospect of escape, that they have come to learn that escape isn't possible. The impact is so significant that even when they have an obvious escape opportunity, they make no attempt to do so. This, sadly, is another state that is all too frequently 'diagnosed' on social media posts with little real evidence.

For the most part, if you are following a good desensitisation and counterconditioning protocol, then reaching a severe case of 'over fear threshold' is unlikely. You will most probably see some warning signs long before that happens. Your dog may change posture, stiffening or leaning into you. You might get some protracted warnings such as low growling or lip-curling. More commonly, they stop responding to verbal cues and will almost certainly stop taking treats. If they do eat, they will grab the food (often uncharacteristically) and then focus back in the direction of the trigger that's provoked their fear. That's your cue to put distance between your dog and the problem, happy-talking them all the way until you reach the distance at which they will take food calmly.

If you take a fearful dog through the gentle, gradual process of desensitisation and counterconditioning and create strong, reliable, positive associations with your presence, you can become a predictable safe base for a fearful dog, gradually introducing new experiences and making sure they stay rewarding. You can never eliminate fear entirely – any more than you can for any of us. Life is never wholly predictable. But your fearful dog can acquire that all-important 'padding' that they never acquired in puppyhood or lost along the way. The fear threshold has moved outwards and they can start to explore.

The idea of safe or secure bases is not new in the field of human psychology. It's well established in the work of clinicians who help to build the relationship (and so the bond) between parents and children. There is also a growing interest in it for doing the same with adults in therapeutic disciplines. As for dogs, you'll have little difficulty finding some well-researched, study-supported literature on dogs acting as safe bases for human dependents. Physical assistance dogs, those who help in day-to-day tasks around the house and the community, can act as a 'security bubble' for their humans. There's also a growing acknowledgement that dogs can offer safe bases to people with emotional health problems. Enter the emotional support dog. If dogs are to perform these tasks for us successfully, then there needs to be an actively sought and maintained 'special' bond between the parties involved.

Despite the plethora of material on the need for safe spaces and secure bases for human-to-human interactions and the service that dogs can perform for us in that regard, there is relatively little discussion of how we can do the same for our canine family members. It's not a completely barren wasteland, but it's not far off. So let's at least start to redress that balance here.

THE FEARFUL DOG

For me, becoming a 'safe base' for your dog is not the inevitable side effect of simply spending time with them, nor does it make up for an absence of training. This is not about training but about building a bond, a relationship, a sense of trust, because if your dog feels safe they can then learn that the wider world is safe. You can unlock the behavioural door that you might have been banging your head against with a fearful dog. (Just to be clear, I use the words 'safe' and 'secure' interchangeably. For our purpose they are intended to mean the same thing.)

I'm going to make plenty of comment about the human-to-human effect of establishing a secure base. That's partly because there has been plenty researched and written, but also to highlight the commonality between the human-to-human bond and the building of relationships in human–dog households. I think you'll pretty quickly get a sense of how much common ground there is between the two.

Let's start at the beginning. What exactly are these safe bases and why are they so important when dealing with behaviour problems in dogs? Well, they're absolutely not about forming an impenetrable shield around your dog through which nothing can pass. The dog needs to learn to be able to interact with the world. Safe bases are 'locations' from which your dog can explore, potentially even out of sight of the base, but with confidence that it is there if it's needed. The safe base is dynamic and might vary in range. It can be static one moment and on the move the next. The safe base can be complex in nature and effect, but easy enough to establish – so much so that it could well be a project assigned to the accidental trainer of Chapter 7. But as casual as the installation of a safe base for a dog might be, it isn't optional. It's as vital as a decent malware programme on your computer, so much so that there is something to be said for making sure – particularly with an anxious dog – that it is installed and well bedded in before you connect to the 'internet' that is the big, wide world.

That doesn't mean you can't start 'mission safe base' while running other established training protocols designed to help fearful or anxious dogs. But it gives you somewhere to look for an explanation if training has stalled and you can't put your finger on why. It might just be that you need to stop and put some more time into building the connection (the roadmap if you will) between your dog and the safe base. Because the safe base is you – the human family.

In occasional cases, this can misfire. Those readers for whom separation anxiety is a very real problem will be familiar (if they have sought professional help) with the ideas behind attachment theory. The problem with separation anxiety is that there's an over-dependence on the attachment figure. In these cases the attachment figure has been established, but has not been able to evolve into a fully functioning secure base that allows for departures and absences because the dog is confident that they will be only temporary in nature.

There is plenty of research looking at whether domestic dogs develop attachment bonds with their owners and whether that attachment bond results in the building of a secure base similar to that between a human parent and child.[2] The happy news is that the research strongly indicates what I'm sure we dog lovers already knew in our hearts: that dogs can and do develop 'special' attachment bonds to the people they live with. Interestingly, it goes further and finds that dogs are more inclined to play, investigate objects and explore an environment if their caregiver is present than if they are left in the same environment with a stranger. Having a safe base, in the form of their caregiver, allows the dog to venture more confidently into the new and unfamiliar. Most (if not all) of the research conducted was with dogs who are sociable with humans in general, but the nature of the secure base effect would seem to suggest that the outcome could be amplified in dogs who are fearful of strangers. The building of a bond with a prospective caregiver for these dogs is even more important, given the benefits in terms of exploration and discovery that the subsequent secure base can bring. The conclusion is that dogs can form attachments to human caregivers in the same way children do and that results in the caregiver becoming a secure base.

This is all fine and dandy. The difficulty comes when the caregiver is still essentially a stranger to the dog, as in the case with new rescues, for example. In these cases attempting to deal with fear of strangers, agoraphobic behaviour or neophobia more generally can feel horribly like swimming against the tide.

How, then, do we establish the attachment bond and the secure base effect so that these dogs can be most helped? Good question. I'm glad you asked. It's the very premise of this book so there would be some awkward silences if you weren't all that interested.

IT'S A TOUGH JOB BUT YOU'LL HAVE TO DO IT

There's no getting past it, if you are the adopter of a shelter or rescue dog who is afraid of strangers, you have definitely pulled the short straw. Sorry. That wasn't very positive, was it? Look on the bright side though, the first bit is the hardest and the end result is a really special, close connection with your dog. You will also have done something really, really cool. You will have given your dog new hope, confidence and a reason to get up in the morning: to see you.

Let's imagine a couple of scenarios.

> **Scenario 1:** You've done a decent thing. You've adopted a dog and you've discovered that, despite all of the indications to the contrary, he is one screwed-up pupster. I know. It's a tough throw of the dice, but there's no good brooding about it. Remember it's not about what should or shouldn't be happening. It's about what *is* happening. So shake off the frustration and despair there, my friend, because this unexpected

and unwanted development is not the end of the story. Oh no, not even close.

Scenario 2: You have a dog that you've had from a puppy. You picked up the little mischief at the age of about seven or eight weeks, fresh from rolling around with her siblings. The breeder has done a good job of socialising the puppies (or at least they've made a start) and so, despite some bewilderment at first, she settled in well; if she could shout: 'Mama, Papa' at you she would. Unfortunately, you were unlucky, and adolescence saw a shift in behaviour and your once cuddly bundle has developed a pathological mistrust of unfamiliar people. Even Auntie Mary, who was viewed as the best tug toy ever last year, got bitten this holiday season. It's a tricky problem to deal with, but you have some stuff going for you.

Your dog (let's call her Patches) is relatively small and so easy enough to control. She's a good learner who enjoys training sessions. Patches has had her teeth on a few unsuspecting victims by now, but the worst injury to date is some light bruising and a scratch on Auntie Mary, probably sustained as a result of Auntie pulling her hand away, dragging it across one of Patches' teeth. All said and done, she's got a soft mouth or what we call in the trade good ABI (acquired bite inhibition). Patches also loves food, any food, but chicken is her favourite. Not a bad list of things going for us there. Not bad at all. But you have something else going for you too and it is no small thing. Patches has never bitten any of the family, only unfamiliar people. (Auntie Mary only visits occasionally and so she doesn't count as family. No offence, Auntie Mary!) Patches loves her people. You are attachment figures and you are a well-established secure base. That means you serve as her security for exploring and as a source of safety in high-stress situations. You are part of the solution, not the problem. You can make with the chicken and start on a training programme that helps Patches recognise that unfamiliar people, yes, even you Auntie Mary, predict good things.

For those people adopting an adult rescue dog, who are in scenario 1, there's a box on that prognostic list that has not been ticked. You *are*, quite literally, strangers. This means there is no secure base available, not as far as the dog is concerned and, to put it bluntly, that's what matters. In the early days, weeks, even months, the best thing the dog can hope for is a safe space: the space under the stairs, squeezed into a space in the kitchen, under a bed or, perhaps behind a television cabinet or a sofa. It is likely that, under those circumstances, getting food down the new arrival might be challenging and getting them to walk on a lead might be even more problematic, assuming you see them at all from one toilet break to the next. Under these circumstances you are the first job. You need to establish the heady heights of 'familiar' then 'trusted'. You need to become the secure base from which this frightened dog can begin to explore the alien world.

Is the suspense killing you? You want to know the secret to successfully establishing a bond and ultimately the coveted secure base, don't you? Ok! Here it is.

> **Spend quality time with your dog!**

It's not quite as simple as that; I know you are going to (and want to) spend quality time with your dog. But I would wager that isn't going to be as straightforward if you are faced with scenario 1. The likelihood in that event is that you'll be lucky if you get an opportunity to even see the poor mite, let alone play a game of fetch with them. When you are faced with lofty advice from the likes of me about making positive associations, but you feel like you've been lobbing cheese or chicken at the poor kid for weeks with very little indication that you are going to even get a tip at the end of it all, it can be frustrating. It would only be natural to start doubting the decision to give the dog a home in the first place. But here's the thing: there is something you can do and it involves only the slightest change to your routine.

If working a structured desensitisation and counter-conditioning plan is 'active' training, then you could think of this emergency option as 'passive' training. It's so passive that it doesn't even feel like you are doing anything at all.

Now, there's no way around it, the tricky bit is at the very beginning, and it involves some waiting. You do need the dog to be out of the safe space they've adopted, whether that's behind a sofa or under a table or whatever, and sharing your space. The first stages of a fear of strangers plan should have done this for you. Failing that, then doing nothing is almost certainly the better option, rather than trying to coax them out from behind whichever piece of furniture they've adopted as their hiding place. They all habituate eventually and come out. They may not look thrilled about sharing the open spaces with you, but hey! you might find that you make some progress before it all stalls and slows down. However, given time your trajectory will improve. Let's look at Sophie from Romania as an example (more of her story is at the end of this chapter). She would come out from behind the sofa and approach under the table to take food. She might even take a look at what her humans, Rory and Diane, were doing in the kitchen, but any sudden noise or movement and back behind the sofa she would go (the subsequent 'hide' could last for days). When she went on medication to reduce her underlying anxiety and make behaviour modification easier, things definitely looked up. She started to be curious about what might be going on in other parts of the house. Up until then, despite having access to the whole house, she had stayed in the open plan part, choosing the kitchen area or behind the sofa.

It's fair to say that I could see more progress than Rory could, and Diane had made more progress than Rory. Again, it's not unusual for the women in a household to bond more readily with a profoundly anxious dog. To say why, I would have to speculate wildly and that isn't my style. What I can say about Diane's nature is that she was forming a secure base for Sophie without being aware – hence the 'passive' tag – but you can make a point of it. That's what we did when holiday time arrived for the family. It was the summer and, as most people would, they had imagined it with Sophie running around on the beach, perhaps having stolen a hat from one of the kids. Sadly, that wasn't going to happen and so I made arrangements to go and live with Sophie for a couple of weeks while they were away. For the first four or five days I made some progress, but I would suffer huge setbacks. Harness or lead work would be going fine until ... it wasn't. The slightest hint of a sudden noise would send her scurrying back behind the sofa and then, despite me being pretty familiar, there was the absence of Diane and Rory to adjust to.

One morning I got a tip-off that there had been a small shift in how Sophie perceived me. There was a loud, percussive bang outside. Sophie hurtled into the kitchen where I was and, instead of cowering in the corner or deciding the sofa was a better option, she came and sat next to me, leaning in and almost peering around my leg. I dared not move. This was the most significant bonding moment we had experienced together and, regardless of the unfortunate trigger, I was not going to squander it. After thirty minutes my legs were giving way and so I slowly slid down the front of the cabinet I had been standing in front of and sat with her. Sophie settled into a lie down and I gently stroked her ears. That was huge. Enormous. 'A small step for …' (you know the rest).

It was now obvious to me that 'training' for harnesses or leads or anything else for that matter was going to take a back seat and that the relationship was going to be front and centre. (Oh, I should have mentioned that we had by now established two beds that had usurped behind the sofa for 'comfy snoozing spot of the year'. One was in the kitchen, later to migrate to the dining area, and one in the living area. Behind the sofa was a distant memory.)

From then on, every time I went in the kitchen and Sophie was there, I would gently roll or throw a toy. Backwards and forwards, backwards and forwards. Sophie would watch; that's a win!

I discovered that wiping the kitchen cabinets and surfaces was a play invite to Sophie, one she was very willing to respond to. She would now play with me and toys and she would request affection. I would take every opportunity to do one of three things. I would engage her in play. If she didn't respond I would see if she were up for a belly tickle or an ear scratch. If not, then I would just sit nearby and read a book or catch up on WhatsApp messages or emails on my phone. It might look like work but every moment of engagement, however basic, was money in the secure base bank account. You become a portable safe space. You make exploration possible because you have formed a kind of safety bubble around you. Attachment theory in practice with dogs. Cool or what?

THE FEARFUL DOG

SOPHIE'S STORY

In the dark, early hours of a December morning, a van pulled up in front of a house in West London. The driver opened the doors and gently placed the passenger into the arms of the resident. Sophie had arrived from Romania. And she was terrified.

Her new owners, Diane, economist, author and Cambridge professor and Rory, former BBC technology correspondent, podcaster and author, had looked forward to her arrival. They'd seen videos of her with her rescuer in Romania and she looked friendly and lively. Their much-loved dog, Cabbage, had died months earlier and they felt ready now to welcome a new family member.

They weren't rescue dog novices. They knew she would need some time to decompress after the three-day journey and settle into her new home. What was unexpected was just how desperately anxious this new dog would prove to be.

On that first day, when they let her out into the garden, she cowered under a garden table and refused to move. Brought back inside, she disappeared behind the nearest sofa and there she stayed, venturing out only at night for food or to toilet on puppy pads. The shadowy images of her brief nocturnal excursions, captured on webcam, were virtually all they saw of her in the first few days.

I made contact after a shout-out on social media, gave them my credentials and got on a Zoom call. Within hours,

Diane and Rory were working on a counterconditioning plan to create positive associations between their presence and really good-tasting nibbles of food. The early signs were positive. Within a few hours, Sophie was cautiously taking small pieces of cheese from Diane's hand. But overnight, prowling the darkened room, she knocked over the Christmas tree. She wasn't injured, but for her, already fearful, the world had literally fallen on top of her. She retreated to her safe space behind the sofa. We were back to square one.

For Rory and Diane, Sophie was more than a dog in one particular respect. Rory has Parkinson's disease and exercise is an essential part of his treatment. Not surprisingly, he had a vision of long strolls through the parklands and towpaths of West London. A dog that wouldn't come out from behind the sofa was a source of anxiety, longing and hope in almost equal measure. The apparently glacial pace of progress added frustration into the mix. So Rory did what a good technology journalist does and shared the highs and lows with a rapidly growing audience on social media under the hashtag #sophiefromromania. Sophie had become a star.

It would be a long process of rehabilitation. Given how high-profile she became, inevitably there were critics of the approach. There were rescue dog owners who said it *shouldn't* take this long, that the sofa *should* be moved so she couldn't hide behind it, that she *should* be put on a lead and taken out for a walk – she'd enjoy it once she was out of the gate. There were a few vets that raged that she *shouldn't* be

THE FEARFUL DOG

here at all: bringing dogs in from Romania was wrong. But she was here and thankfully, unaware of all the should haves, shouldn't haves, oughts and ought nots. We carried on with the job of growing her bond with her people to provide an all-important anchor as we gently extended her range of experiences.

I encouraged Rory and Diane to keep building positive associations at every opportunity, casually dropping treats for Sophie near the mouth of her selected 'den' or when she poked her cautious face out from behind the sofa. They could do absolutely nothing but wait for Sophie to surface and habituate to both the house and them.

The knack with these things is to always go at the dog's pace. Resist the temptation to try and coax or persuade them to come closer when their body language indicates they really don't want to. The tip-off is how close they come towards you of their own free will before stopping. That's the 'under the fear threshold' limit and it's always good to pay heed to it. Things will progress a lot quicker if you do. The tell is often the stretch. If you're holding the food in your hand, or you've thrown it a little too short and the dog has to stretch to reach it, they can be conflicted about what risk to take. Finally, they grab the food and run for cover, away from you. Your next opportunity will come later, often much later, than it would have had you bided your time and trusted the process.

And so it was with Sophie. Day in, day out. As time went by, she would come out and take food from closer to her people. Her favourite place to do that was from under the table, although she would sometimes cautiously approach either Diane or Rory at the very sofa that she would spend much of her time behind. In the early weeks and months there was little chance of being able to touch Sophie without panicking her. Indeed, even standing up from sitting would be enough to have her heading for the safety of her sofa.

It was tough for all concerned. Life has to go on when you have work and family commitments. You can modify them, for sure, and Diane and Rory did an awful lot of that. But it was hard for Sophie too: new people, new sounds, new smells, a new place. Nothing was familiar and it's not unusual for dogs like Sophie to find the new very difficult to deal with. So, this was a marathon, not a sprint, to coin an overused phrase. A few steps forward were followed by a step back, followed by a plateau of hours, days, weeks, then a huge leap forward and another plateau or a drop back. It's not work for the faint-hearted or the chronically impatient. Rory, by his own admission, is in the latter camp, but, if you ask me, he's acquired the skill of patience. Sophie didn't give him much choice.

'IT DOESN'T FIX THE PROBLEM, BUT IT MAKES IT FIXABLE.'

We'll talk more about medical interventions in a later chapter but, for now, it's enough to say that there came a point when it was clear that, given Sophie's continued high degree of anxiety, medication might help. After a discussion and consultation with her vets, they agreed to put her on a selective serotonin reuptake inhibitor called Fluoxetine – essentially 'Prozac for dogs'. It's important to stress here that medications such as SSRIs and TCAs (tricyclic antidepressants) are not intended to fix or cure the problem, but they can and do make it easier to carry through a programme of behaviour modification. If you reduce the underlying anxiety by giving medication, you can increase engagement significantly. That proved to be the case for Sophie.

There's a tendency in modern society to see medication for deep-seated anxiety as a last resort. There's still, to a degree, a sense that such interventions are mind-altering, causing 'stupor' and taking away the intrinsic personality of the subject. It's not the case with modern medicine and I would argue that far from being a solution of last resort, anti-anxiety medications should be a front-line consideration for dogs who find the human world hard to cope with. If it improves their quality of life and helps them to make the connections that will give them a shot at happiness, why wouldn't you?

Sophie began to make more progress. She learned to play and delighted in a rough and tumble with her people and her toys. Play, in fact, was definitely the way to bond with Sophie. She adopted the comfortable dog beds that had been available since day one and abandoned her safe place behind the sofa. She would hang out with Rory in the living room as he watched the football or lie down next to him when he played the piano. She joyfully joined Diane when she did floor exercises to strengthen her back, mirroring her stretches. She gave all the appearance of weighing up the merits of *Strictly Come Dancing* contestants on autumn Saturday evenings. She roared enthusiastically around the garden after pigeons.

But she remained extremely wary of new people and had to be introduced to them gradually and carefully over time. And after a year, Rory and Diane still struggled to accustom her to a lead or harness. The walk remained elusive.

But with dogs like Sophie breakthroughs do come. After a month of intensive work and, more importantly, playing with her, I finally succeeded in getting her to offer her head through the harness and clipped it on. A whole world of possibilities opened up. Finally, a first walk.

Sophie is still a 'work in progress'. At the time of writing, she remains anxious about strange people, strange events and noises. She's showing signs of being interested in other dogs and has undoubtedly discovered the joys of play with her familiar people. Her social network is expanding slowly and she happily wears her harness to explore the park.

THE FEARFUL DOG

Nothing about dealing with profoundly fearful dogs is quick or easy. If you're in that place and you're working through it, then you have my admiration. The brutal truth about fear is that you just don't know how long it will take or how far down the road to recovery you're going to get. Do your job and do it well and you'll get some of the way, possibly all of the way, but you can't put a timescale on it. There are no 'should haves'. The dog sets the pace.

What Sophie has done for me is remind me of just how valuable play and a gentle focus on the dog's needs and wants are. They may be deeply fearful of a lot of things but being attentive to when they are willing to engage with you, or have you engage with them, will help them to develop a meaningful bond with you.

If a dog doesn't feel safe, they are not likely to learn the things you intend for them to learn. Safety breeds positive emotional responses to events and places, but most of all to you.

That's what is meant by a secure base: a dependable person that they can use as a safe reference point. People become anchor points in an environment they find uncertain. Alongside them, they can explore and discover the world. There is a contract between you, the human, and the dog. That contract is to never let anything bad happen to them and to be there as a shield if, by no fault of anyone, it does. It's an idea, or an understanding that has been accepted and harnessed in human psychology for some time to help anxious or traumatised children and adults alike. It's a concept that the dog world would do well to consider more often. It became very much a focus for me to establish Diane and Rory as Sophie's secure bases. For Rory particularly, it perhaps took the pressure off for both of them. It took the focus off needing Sophie to do things and put it on doing things with Sophie. It puts the emphasis on the bond, back on that ancient connection between people and their dogs.

CHAPTER 11

THE REACTIVE DOG

I spent some time considering the right title for this chapter. Should it be 'The reactive dog' or should it be 'The aggressive dog'? In some sense neither are right or adequate. To some extent, using the word 'reactive' when dealing with dogs who are explosive in their behaviour is understandable. Given the wide interpretation of the word 'aggressive' in modern society, I can see why many would want to avoid it in order to defend their dog. 'Reactive' seems somehow gentler, shorter in duration, something that has a trigger that justifies the behaviour. In many ways, that is true of the reactive or aggressive dog. There is no such thing as a motiveless action in animal behaviour. And therein lies the rub: aggression, motive, intention.

For many of us 'aggressive' has connotations of violence or, more precisely, violent crime. If our dogs are seen or described as aggressive, it seems that we're finding them guilty of an egregious, unwarranted act that has hurt a third party for personal gain or satisfaction. It may be 'motiveless' or, worse, malicious – I could go on. So no wonder dog guardians would rather find a more pertinent, less loaded word to describe how their dogs are. However, there's a flaw in the language whichever way this cookie crumbles. No matter what vocabulary you use, the world will form their own conclusions. How many people with a 'reactive dog' have heard the accusation 'That's an aggressive dog!' Plenty, I would imagine. Certainly I've heard from enough clients who have had that hurled at them to reasonably conclude that it's a widespread viewpoint (though I don't have research figures on it).

The reality, however, is much more nuanced. Neither 'reactive' nor 'aggressive' are character descriptions, a summing up of all that they are or will be. And at this stage, by way of making a point, I'm going to drop the 'reactive' tag and stubbornly push for the reinstatement of 'aggression' as a more meaningful way of thinking about a dog's behaviour. It implies no value judgement, it is not a label to hang around a dog's neck, it should not be a source of shame or guilt and it certainly does need some serious unpacking.

Aggression describes a set of behaviours that have a purpose. For animals, this is more straightforward than for humans. The purpose, in its widest sense, is to make something stop or to go further away. A dog who bares their teeth or growls or even bites isn't an aggressive dog; they are a dog who behaves aggressively (or exhibits aggressive behaviour) in a specific context. It's never random, never out of the blue and never without provocation. All behaviour, violent or not, is 'provoked'. They all have an antecedent or something that prompts the behaviour. Whether you intended to provoke the big dog with sharp pointy teeth to bite you or not is wholly irrelevant. Rather than focusing on the 'crime', we really should be focusing on the motivation and the mitigating circumstances. In just about every situation I can think of where a dog is concerned, there isn't a time where the action hasn't been explainable and yes, as far as the dog is concerned, justified. That's a bold statement, I know, but think about it for a moment. Your dog has limited options for getting out of a situation that actively threatens themselves or a valued resource: fight, flight or freeze. There might be some degree of genetic preloading in this regard, but the impact of that genetic predisposition will soon be escalated if proved, through experience, to be successful. Ultimately what initially begins as a spontaneous response becomes a learned one, maintained by virtue of repeated success. Therefore it's our responsibility, both to our dogs and to the public, to expand the range of options our dogs have in navigating the world, either by reducing the likelihood of the threatening trigger or, preferably, by changing the dog's view of the trigger from seeing dragons and demons to one where they anticipate reward.

I'm not being hippy-dippy or trying to minimise a really serious issue here. Of course, the unavoidable limitations in a dog's choices of action when threatened can make the 'upset' dog potentially dangerous and it is no defence in a court of law to say, 'I know why he did it, m'lud, and frankly I don't blame him.' Indeed, that kind of thing is likely to get you banged up at the same time as your dog. The difference is you are likely to re-emerge from your incarceration; your dog may not get the same opportunity. In other words, we all have a duty under the law to keep society safe to the best of our ability and, sadly, that sometimes means acknowledging that the risk of danger and our ability to mitigate against it is just too much. But let's put the language issue to bed. I use the words 'reactive' and 'aggressive' interchangeably throughout this chapter; for accuracy, that might include the term 'reacted in an aggressive manner'.

Before everything becomes too gloomy, let us take a look at the options we have with dogs who snarl, bark and bite. All is a very long way from lost.

DON'T STAND TOO CLOSE TO ME

Proximity sensitivity is, I suspect, something that many people who have reactive dogs will recognise. They'll report that their dog is fine, provided the other dog or person doesn't get in their face. That's when they go off and go off big time. It can be deceiving and people can be easily caught out by it. Because their dog seems fine for much of the time, they live in the neverending, but forlorn hope that the next time will be OK, that from then on their dog will discover the value of friendly relations and up-close encounters. Unfortunately, this is very rarely the case and allowing repeated approaches may only exacerbate the problem.

What's really going on? These dogs are tricky to read. For the most part, the presence of the problem (the other dog or person) seems neutral. Your dog may even look curious. But there's a booby trap waiting for the poor unsuspecting human, eternally hoping for peace to break out. The circumstances may seem all too familiar to dogs with proximity sensitivity, but as far as the dog is concerned it's familiarity — or rather the lack of it — that's the problem. The absence of familiarity means the fuse wire is lit and sparks its way towards the near inevitable powder keg of doom. In truth, it's rarely that bad, but for the human at the other end of the leash, it can feel so. Dogs who suffer from proximity sensitivity are no different from those who make their discomfort clearly felt at forty yards. The only difference is how they deal with the exposure over distance. Many dogs who react in an aggressive manner towards scary things are no-nonsense characters. 'Let's get that scary thing out of here right now. Circle the wagons. Throw everything at 'em. By jingo, they're not getting a yard closer.'

There seems little ambiguity there. (Actually there is but we'll come to that a little later.) Your proximity sensitive dog is another matter altogether. They behave as if the presence of the problem is fine. They even appear relaxed and nonchalant. You could, indeed, be forgiven for thinking all was well and wait for the mutual rear end sniff and first date kiss. Instead, at the first nose-to-nose possibility (sometimes sooner), the proximity sensitive perpetrator opens fire. The victim either retreats, their feelings hurt, or returns fire and a fight breaks out. Depending on the fight styles of the two dogs that might end quickly with no injury to either party, or intervention might be needed from the handlers.

Now, as much as I said at the beginning that this isn't a 'how to' book, I'm going to give a little advice here because it's important. The law of averages dictates that at least some of you reading this will have been in this situation at some time or another. A few of you will have got in the middle of things or grabbed your dog's collar and you will have got away with it. No injury to you or your dog. All good. General mutual apologies expressed and everyone humbly backs away from the situation. There's a strong chance, however, that if you have got between the dogs or put your hand in and grabbed a collar that you will be the one who comes off worse from the altercation. You will have picked up a redirected bite — a bite not specifically intended for you, but you were the only thing that your dog could reach to offload the pile of adrenaline and cortisol coursing through their body. It's a tough gig and, as you slump home feeling wounded in more ways than just physically, you might even have told your dog off for humiliating you in public. That's understandable and you can forgive yourself if that is the case. Getting angry is a bad plan for reasons that I will come to shortly but, for now, forgive yourself. We are only animals and dog fights scare us so, of course, we might react in an aggressive manner ourselves when they happen. Some of you might even have been on the receiving end of the other person's anger at the situation, despite your effusive apology. It's the same thing. They are adrenaline-pumped and scared too. It's only natural but we can train ourselves to take a different beat.

Back to the lesson. If the fight has persisted for a few seconds or more (intervention delay is going to be a matter of personal bias) and noise or water hasn't stopped it (a dog fight is about the only time when a mild aversive such as noise or throwing water is justified), decide who is the likely perpetrator or instigator of the fight and who is the victim. Take hold of the perpetrator, either at the base of the tail or under their back legs where they meet the body, and lift swiftly until the dog is almost vertical. This is called wheelbarrowing. Swivel away from the action and then let go of the dog. As a rule, they will not re-engage with the fight; if they do, simply repeat and get them on a lead when all four of their paws are on the floor. Why do this? For one thing, it generally takes the heat out of the situation, increases distance (which for the perpetrator, at least, is the right thing to do) and, most importantly, they can't deliver that embarrassing and painful redirected bite we were talking about earlier. This is because in order to bite you they have to lift a front leg off the floor which, because they only have their front paws on the floor, destabilises them significantly. Do all of this swiftly and decisively. In this case fortune favours the brave.

As different as they may look, these two dogs who've got themselves into a spat are more alike than you might imagine. The proximity sensitive dog isn't 'fine' at all. They are only just holding it together until, when close to touching, they can't handle it anymore; given that feigning disinterest hasn't worked, they've gone full-on guns blazing. If these encounters are allowed to repeat themselves, then a proximity sensitive dog may learn that faking disinterest is a pointless exercise and things will get more violent sooner in the encounter. It's important to recognise this when it happens and resolve to employ a desensitisation and counterconditioning protocol as you would with any fearful dog. If necessary, start a long way back. There is no harm in overcompensating. It's all emotional money in the emotional bank account. You'll occasionally hear people talk about 'dogs sorting it out among themselves'. They may well do so, but not in a way that's helpful or safe for you or them.

A word about aversive techniques or reactions on the human's part here and the language I've used. While I have described the participating dogs as 'perpetrators and victims', that isn't really the dynamic here. It's largely about two frightened dogs and there is no 'guilt' to be ascribed.

By aversive techniques I mean anything that you might deploy to discourage the dog and make it 'think twice' the next time that situation arises. To work, it will need to be unpleasant, delivered with timing accurate enough that the dog connects it with the 'offence' and severe enough for the dog to want to avoid or escape it in the future. And this is problematic on a number of counts. The first count (and I would argue the big one) is that the negative emotional response the dog already has to the stimulus (in the case outlined, the close proximity of another dog) is not going to get any better. You might have stopped the aggressive displays, but you have not changed the emotion. Next time you walk up to a frightening stimulus, the dog doesn't go off at any distance, but you've broken the unspoken contract that says, 'I'll keep you away from scary things'. The alternatives the dog has for dealing with the problem have reduced. If they perceive the threat to be powerful enough, where you may not have experienced biting in the past, you might very well now. You have punished out, not the negative emotional response, but the message they have been trying to give you that there's a problem. That's bad news.

For proximity sensitive dogs, the name of the game is brief encounters: learning to read the dog's body language so that you know what's too close or too long a contact for their comfort and moving them swiftly away before the reaction erupts or the spat starts.

F.O.D. (FEAR OF DOGS)

Fear of dogs registers high on the list of the most distressing problems that dog guardians face. Possibly only usurped by fear of people (or stranger danger) and the heartbreaking and restricting separation anxiety.

For those dogs who are not candidates for proximity sensitivity, fear of dogs presents as highly explosive, aggressive responses to other dogs, both on and off lead. These responses may or may not result in fighting but certainly looks dramatic and violent. In these cases, the priority with any training protocol is to change the emotional response of the dog. They don't like other dogs, so the purpose of their explosive reaction is simply to get the scary dog to go away. This is, in the first instance, a spontaneous response to a problem being present. Over time, the dog learns that it is effective, which maintains the behaviour. Now, the outcome of the aggressive reaction to the problem dog might be that the handler turns away and puts more distance between them and the problem. The result is hopefully enough distance that the dog is put back under their fear threshold and will accept food. I suspect that there are some of you, understandably, shouting at this moment: 'Accept food? But aren't you reinforcing the bad behaviour?' But if the motivation for reactive or aggressive behaviour is fear, that isn't the case. (I'll explain more about this later.) In fact, for a dog that fears other dogs, the 'turn and go' achieves the desired goal for the dog, which is to increase distance so the dog has no need for a continued display of aggression.

Do things right and a well-structured desensitisation and counterconditioning protocol will, over time, gradually reduce the distance from the other dog that the fearful one feels to be safe.

F.O.P. (FEAR OF PEOPLE)

It's hard to imagine a behaviour problem that is going to impact people's lives quite as much as this one. It's particularly true if your dog's strategy for avoiding or escaping strange people includes aggressive behaviour. It affects the time of day you might walk your dog, the walking routes you take and when, if at all, you have visitors to the house. It can have serious legal ramifications if your dog is loose and bites a person or, indeed, if they bite a visitor to the house. The serious consequences that might result from a handling error is why this, above all other behaviour issues, should involve professional help and why your chosen professional should ask you some important questions around your dog's bite history and chosen response to strangers. They should also require you to train your dog to willingly accept wearing a muzzle if it proves necessary. From there on, the solution is the same as for all fear-related issues. A considered desensitisation and counter-conditioning plan is needed to create association between people and something the dog really likes. The trainer can also train an alternative behaviour with the primary intention of achieving a classical conditioning side effect, where the dog acquires a positive conditioned emotional response to strangers. If we think back to the Pavlovian 'A' predicts 'B' approach to creating a positive association, training an alternative behaviour works this way: The dog sees someone who worries them (A). They are asked for a behaviour that they know well and already perform reliably, they do this behaviour and get a treat (B). So, as far as they are concerned, the person (A) predicts being asked for the behaviour, which predicts the treat (B). Ergo, the person predicts something positive. Both approaches will always require good timing on the part of the trainer, as well as an ability to recognise a positive conditioned emotional response when they see one. This is important because it will avoid the most common error in this training process, which is pushing to the next level too early.

Just as it was starting to sound straightforward. Typical!

'I'M NOT SCARED, YOU'RE JUST IN MY WAY!'

There is a twist in the tail to all of this. Some of you might have spotted the tip-off in the earlier sections. There's a reference to the behaviour of the subject dog *both on and off lead*. If you spotted that and it caused you to raise an eyebrow, you get a bonus point. Here's the thing: the two most common reasons for a dog to react aggressively to a stimulus when on lead, such as another dog or person, are fear and frustration. You couldn't get two more diametrically opposed motivators for a behaviour and yet those two things look almost identical. Now we have a problem. One of these motivations, fear, is for the scary thing to go away; to increase distance because the fearful dog wants to avoid contact. That's straightforward enough, I think we would all agree.

The problem is that the other motivator, frustration, is triggered because a 'not upset', unafraid and, in fact, quite socially minded dog wants to *decrease* distance so they can go over there, say 'hi' and investigate, but can't because the lead is slowing or stopping their approach. The ability to greet and investigate in a way that is fine and appropriate for both the subject dog and the stimulus (the person or other dog) is hindered. (Gates and fences act in a similar way, which is why you'll get dogs barking up a storm when they're behind them and see other dogs or people.) As a result, frustration builds up and if or when the dog eventually does reach the object of their desire, they are so amped up that they snap or, in the case of meeting another social dog, they squabble, leaving the handler in the misguided belief that their dog isn't sociable when the opposite is true. It's kind of like road rage. When drivers are stuck in a traffic jam and can't get where they want to go, some of them will start seething, beeping their horn and might then give a full-on aggressive display at the next person who comes within shouting and fist-waving distance. In some ways, we humans are not all that different.

For a frustrated dog, the reward is to go and meet the other dog. They are friendly but the lead restricts their options and so they get frustrated if the lead goes tight. Of course, this can be a self-fulfilling prophecy. Their excitement and eagerness to investigate brings them to the end of the lead in short order. So turning them away and putting more distance between them and the object of their desire acts as a 'fine' for the reactive behaviour. They learn over subsequent turns that the way to meet is to stay calm and approach politely. If you want to go over to say 'hi', you have to arrive in a civilised fashion.

THE 'REWARDING BAD BEHAVIOUR' MISCONCEPTION

I promised to come back to this conundrum. Why do I encourage dog guardians to feed their fearful dog, regardless of whether they are going off or not? Surely you are reinforcing the 'bad' behaviour? Isn't that just confirming to the dog that they were right in their aggressive response and aren't you just going to get more of the same as a result?

The answer is 'No!' Good, I'm glad we cleared that up. Let's move on ... just kidding! Let's address this because it's a common misconception and it's important to correct it so that you don't miss opportunities to make those all-important positive associations that are so needed by fearful dogs.

Reward or reinforcement are terms we use for dogs who are not upset. It's the consequence for actions that the dog chooses to take in order to access something nice, such as chicken or sausage. If something they choose to do has a positive or good outcome, then they are likely to do it more frequently. If they do, then you can say that behaviour has been reinforced or rewarded. The secret is in the motivation. A dog performing tricks to access reinforcement is motivated by access to the pay. A dog motivated by fear is looking to escape or avoid something frightening that may even be life-threatening. At that point that dog isn't concerned with accessing the chicken in your treat pouch.

Self-preservation is the priority. But the chicken can play a part – not to change behaviour directly, but to change the dog's emotional response to what they perceive as scary: scary thing means chicken. Eventually, the reactive or aggressive behaviour becomes redundant because the stimulus is no longer scary. It's a matter of contingencies. For 'not upset' dogs the pay is subject to them performing the right behaviour. For 'upset' dogs the pay is only subject to the scary thing being there.

STEP AWAY FROM MY STUFF

The world of behaviour science is complex enough without adding to that complexity by sub-categorising everything until we disappear into a black hole of ever-increasing variability. (That's my view anyway.) So, keep it simple is my mantra – or at least as simple as you can. That's why I tend to lump aggressive behaviour on the 'upset' side of things into only two camps: the fear of harm to yourself and the fear that you're going to lose something that is valuable to you. That mostly covers it. The big hitter there, and the one that pretty much trumps everything else, is fear of harm to yourself. For dogs, the idea that they are either going to be seriously injured or even killed is going to make everything else pale into insignificance. That fear can be really powerful, which makes it difficult for us as trainers to predict how far we'll get in resolving the issue. You will always get some of the way, for sure. You might even get all of the way and end up with one of the community's leading socialites, you never know. With clients with F.O.D. I can say with some confidence that I will get them walking past other dogs at a road width apart, but whether they will ever greet or play with another dog is something that only trends and trajectories over time are going to tell you, so I tend to lean in favour of managing expectations at the beginning. That way there might be a pleasant surprise or two in store further down the line, which is always better than disappointment.

For those people who find that they have taken a resource guarder into their home, then there is some good news. If you had to have one problem behaviour in your dog, then resource guarding is the one you want. And of all the resources you want them to guard, be it food, objects, locations or people, food is the easiest to modify, as a rule, although they are all capable of being resolved, given the right training.

It's important to be clear that the objective with any behaviour problem is never to see the problem again. Life being what it is, however, probability dictates that you are going to make the odd mistake, or there will be minor regressions and the behaviour will pop out and say 'boo' once in a while. But it isn't true that you have to trigger the behaviour to fix it; in fact, quite the reverse is true.

If you get to the end of a resource guarding plan and your dog hasn't guarded even a teensy bit from start to finish and you are not a professional trainer, then my hat goes off to you. And, I might add, have you considered a career in dog training?

When it comes to changing the behaviour, resource guarding is less of a challenge than fear of dogs or people because it isn't about *personal* safety but rather the safety of their *stuff*. It's not a life-or-death situation. I might even define resource guarding as anxiety, rather than a fear. All that said, it can still lead to humans who pose a threat to the valued resources getting bitten and, particularly if there are kids around, that's no small issue. The principles for dealing with fear hold good for anxieties like resource guarding. We're looking to make positive associations between the problem and something they already like. We want to take the heat out of any situation and we want to show that any action on our part is not a threat to their valued resources or an indicator of bad stuff on the horizon. The best way we are going to do that is to recognise that this isn't bad behaviour. It might be unwanted, but it isn't 'bad'. It comes from a place of fear or anxiety. We need to recognise that and offer patience and kindness if we are going to help our 'reactive dogs' in a way that is good both for them and the society in which they are required to live.

THE REACTIVE HUMAN

We could be easily swayed from showing that patience and kindness in modern life. We live in a polarised and troubling world. It feels like everywhere we look there is division, contradiction and confusion: the cost of living crisis, the NHS crisis, the climate crisis – not to mention a couple of years of pandemic isolation feeding anxieties and uncertainties and a few ugly wars thrown in for good measure. There are politicians promising the earth with no plan for how they might deliver it, but utterly convinced that it's the other party that's responsible for all the wrongs in our lives. On social media it seems as if everyone is calling everyone else out for a fight. It's hardly any wonder that people are stressed and anxious. And on top of all that, the dog is playing up. Snarking (a mix of snarling and barking) the friendly dog next door and the neighbour, despite the fact that you and they have lived there for aeons. He's even growling at you (your dog, not the neighbour) even though you are the one that puts the food in his bowl, turns out every day for walkies even if it's absolutely chucking it down with rain. You put a roof over his head. It's not like he has to live in a kennel outside. The ingratitude of it all!

Obviously, I want to help alleviate any stress that your dog's 'problem' behaviour may be causing you, but first I want to acknowledge that it's understandable that you respond in a negative way to it. You're not alone. Just like your dog, you're reacting to circumstances that you find worrying. Instead of getting stressed, I want to give you a way of thinking about your dog that means you just need to follow a system to make real progress. A way of thinking that helps you to understand what is going on and how your dog uses behaviour (even the violent kind) to communicate a serious need.

It's easy to be tempted by promises of 'quick fixes' from people on Facebook or TikTok. After all, the promise sounds plausible, right? It's an enticing argument. 'That's how pack animals do things.' 'Dogs need firm leaders. Powerful ones.' 'A healthy mix of kindness and discipline.' 'How are they going to learn to deal with the tough things in life if you pussyfoot around them?' 'Teach them who's boss.' Sound familiar? There was a time when everyone believed that the slap of a 'firm hand' was all that was needed to keep children on the straight and narrow. We've moved on from that as a society, so let's not revert to this as a way to treat our dogs, those family members who show joy whenever we return, whether it's from a two-week trip away or two minutes in the loo. As seductive as the promise of a quick fix may feel when you're lost for what to do, just turn away. I explain often enough in this book why using aversive or coercive methods is a bad idea, so I won't press the matter again here. It's enough to know that snake oil never healed anything. My guess is that if you have gotten this far through this book then you have already drawn that conclusion and are looking for something else. Which is why I'm here.

Force-free or positive reinforcement training has long been supported by study, research and scientific scrutiny as the most effective way of helping dogs. Any dogs. That goes for cuddly, friendly, taking-anything-in-their-stride pooches to scared, cowering dogs or those who behave in an aggressive way when faced with a scary stimulus and, for the record, that includes dogs who bite. (Although dogs that have caused serious harm are unlikely to get as far as a trainer or behaviourist as it's likely the law has already intervened.) While no one with any credibility can say that they can modify a hard mouth (i.e. one that bites), what they can say is that they can significantly reduce the risk that the dog will use it. If you hold the line and see things through, then the relationship can be mended. Household harmony can be restored.

There's that 'relationship' word again. That's because I really do believe that is the work I'm in: relationship building and relationship mending. This isn't hippy-dippy, happy-clappy stuff: the relationship people have with their dogs is important. In the 2023 PDSA Animal Wellbeing (PAW) Report,[1] the top three reasons for getting a dog were: love and affection, companionship and to complete the family. That's a big emotional job our dogs are doing. When there's a problem, it can significantly affect our quality of our life, as well as theirs. So, at this point, I want to give a big shout-out to all those professionals in the dog behaviour world who see the people in the family as deserving of help as much as the dog. While dog training may seem like a good fit for those who prefer animal company to that of their fellow humans, in truth it isn't. If you don't like people (and I make no judgement of that) dog training is, in fact, not the thing for you. I spend as much time in consultations with people looking for the 'aha' moment – that point when what's really happening dawns on them – as I demo training their dog or I'm explaining how to train.

It's always about how dogs learn and what motivates them. Once people understand that they can start connecting with their dog again and that's the return to 'us' time, play time, bonding time.

WINSTON'S STORY

I can safely say that every dog I've helped and trained during my career has touched me deeply. They all imprint on my heart like they've inked the rubber stamp and pressed it firmly in place. I remember them all and, with some, I've been lucky enough to build lasting attachments. One of these was Winston. He was nothing short of majestic with a deep, glossy, chestnut-red coat, rippling with muscle. A Rhodesian Ridgeback, he was mountainous in size. He was the biggest Ridgeback I had ever seen, which always made Iola (his devoted mum) laugh. Having grown up in South Africa, she was used to the scale of these 'Lion fighters', which were what working Ridgebacks were back in the day. If he set his bin-lid sized paws on your shoulders, he could quite easily whisper in your ear. The problem was that he wasn't really a whispering kind of dog.

When Iola contacted me for help, she was living in an idyllic cottage in the Cotswolds with Winston and his housemate, Ruby the Dachshund. Life should have been good; instead, it was becoming increasingly stressful. Winston was beginning to scare some in the community. He was reactive on lead to dogs and to quite a few of the people he encountered. He didn't much care for horses either, which pretty much ruled out most of the Cotswolds inhabitants they were likely to regularly come across on a walk. Ruby, on the other hand, he adored. She and Iola were his safe haven.

I came armed to that first meeting with the full treat toolkit of chicken, frankfurters and beef steak, all the seriously big hitters in the reward stakes. As Winston climbed the rickety gate to check me out, I was glad I hadn't come empty-handed. This was going to be trade negotiation at its most intense.

But fortune favours the brave, particularly those who come laden with a bag of good stuff. It didn't take me long to get on Winston's good side, which was very promising. It boded well, at least in terms of the fear of strange people; I was, after all, a stranger, albeit one bearing gifts.

Pretty soon both Winston and Ruby were demonstrating their keenness to learn, and Iola and I had started to formulate a plan for Winston's issue with passers-by and dogs when out on walks. The first challenge was how to control a dog that had considerable weight and muscle on his side if it were to come to a contest of wills about direction of travel. We worked on close heel and he caught on quickly. Being food-motivated helped considerably. But he had by now, shall we say, developed a reputation among some in the neighbourhood. I noticed on a subsequent visit that the rickety gate had been replaced by the kind of heavy-swinging, none-shall-pass barrier that you might imagine a giant living behind.

The reality with Winston was that he was a small, anxious dog in a big dog's body. There is no doubt that he was alarming when he reacted to dogs or people because he was huge and he was loud. He acted and sounded like a drunken prize fighter, calling out to passers-by, challenging all comers to be man or dog enough to take him on. In fact, the last thing he wanted was a fight. He reacted because he was afraid. If he could make enough of a display, he could get all these scary things to back off and go away. Whereas a smaller dog might have been cut some slack by others that he encountered out on walks, because of his size, there was little chance of this with Winston. We could hardly blame them. We had to make him see the world in a different way, one that he was willing to engage with, rather than roar at to go away.

There were times when Iola worried that, however much she loved him, Winston's reactivity would be too much for her to handle. However, thanks to her determination not to give up, he made great strides. He learned through the kind application of classical conditioning, making positive associations with things that scared him that the world was essentially a safe place and, as a rule, most people and (at least) some dogs were not a danger to him. I remember Winston as affectionate, gorgeous and goofy. When I got to know him, I stopped seeing his size. He always seemed just a little fella with an expressive brow, which he would crease when puzzled about something and who liked life to be predictable. He was happiest when he could anticipate what was likely to happen next.

When we built a home office cabin in the garden of our home in Wiltshire to offer intensive one-dog-at-a-time board and training, Winston was one of our first guests. The progress that Iola had already made was clear to me, though it had probably crept up without her completely noticing. Still, a couple of weeks away would give her a break and I could work with him intensively. I found locations with a steady flow of dog and people traffic, but plenty of open space where we could create distance if he showed any sign of being overwhelmed. Every sight of another dog, every step-by-step closer approach was rewarded. We became well-known to the regular dog walkers, some of whom became quite invested in our training and would wave or give thumbs up.

Being just a small dog in the wrong-sized body, Winston liked other small dogs. He was desperate to be pals with Ripley. Ripley was less convinced. What she saw was a giant and one that wasn't all that conscious of what he was doing with his body and where he was putting his paws. But, over time, she grew accustomed and would lead the way with him on walks.

He came to stay with us again a few months later. Iola had long wanted another Dachshund and had decided that she would let Ruby have a litter and keep one as a new member of the family. The whole build-up to the event had been increasingly stressful for Winston and when the puppies arrived and he caught a brief glimpse, the poor lad was terrified. This was new in the place he felt safe and new was very, very difficult for Winston.

Iola emailed us: Could we take Winston so that Ruby could relax with her puppies – and so could the humans? Of course, we could. I went to collect him and settled him into the cabin. Such was Winston's way that even if invited to spend time in the main house, he wouldn't stay long. A brief visit was enough and then he'd head back to the cabin and would be found patiently sitting outside, waiting to be let in. From that point on, the cabin became known as 'Winston's holiday cottage'.

Back home, Winston continued to make real progress. People in the local dog-walking community opened up to his charms and he found some dog chums that he could walk with confidently. An experienced dog walker helped expand his social circle still further.

We moved to Scotland and, though we'd kept in touch, we hadn't seen Winston for a few years. Then Iola sent a message. Winston had died. She was devastated. He'd been in boarding kennels when she had gone away for a short holiday and had been struck down in the night with gastric torsion, a condition that can affect large, deep-chested dogs as they get older. I was rocked by a wave of aching sadness. Winston with his oversized paws and puzzled frown, who'd worked so hard to learn to cope with the world. It was so heartbreakingly unfair. He was only a little dog.

CHAPTER 12

GREAT EXPECTATIONS: THE RIGHT DOG

You can't choose your relatives, but you can choose your dog and it's an important decision. There's a lot of rational advice given around the subject: don't just decide on looks, choosing a rescue is the responsible thing to do, do your breed homework, don't be influenced by fads and fashions. All of which is quite right. There's also, sadly, plenty of personal prejudice broadcast as fact, and lofty, unequivocal assessments about the suitability of different types of dogs for different types of people made by sections of the animal professional community who really should know better. Put out on social media, scraped and republished in newsfeeds, like the great dog myths, they start to gain currency through sheer weight of repetition. You'll know the sort of thing I mean: 'I'm a dog trainer and this is the breed a first-time owner should NEVER get.' This was a Cocker Spaniel, by the way. I mean, really? Or 'I'm a vet and I'd never get this type of dog, they're all bonkers!' Doodles, in this case. And, yes, they can be bonkers – deliciously, delightfully so, but I've also met Doodles who are relaxed and chilled. All of this means that doing the due diligence research into what might be the right dog for you leads rapidly into a quagmire of contradictory advice that can be head-spinningly hard to navigate.

Of course, in relationships, pure reason is – let's face it – not often the absolute driver of decisions. You're going to be spending, hopefully, a good chunk of your life with your dog. It may not be as intimate a relationship as with a human partner, but emotionally, it can be pretty darn close and sometimes longer-lasting. So whether your heart is being tugged towards a glossy pedigree or a wonky, overlooked shelter mutt, there's plenty of emotion at play. So let's acknowledge and celebrate that from the get-go and then think about basing an emotional choice on solid ground.

However much we might be prone to puppy love at first sight, since we want to avoid an early relationship breakdown, there are some practical considerations that are worth pondering: age and size, cost to feed and insure, health issues, energy levels, relative strength, as well as the lifestyle that your new friend will be joining. Are you looking to run or hike for miles or just have a gentle ramble to the pub and sit with a pint? Once upon a time and not that long a time ago, if you worked for a living you were expected to resign yourself to a dogless life. These days, with all the available support service businesses, that need not be the case, but you'll still want to have the time and energy to spend your leisure hours training, exercising and generally hanging out with your mate. As most dog lovers know, they do have a talent for getting you to bend your life around them – and rightly so.

THE PULL OF THE PUREBREED

When it comes to thinking about different breeds of dogs, it's worth repeating that dogs are individuals. No matter how diligently we compare the temperament descriptors that are tagged on to different breeds, they aren't a guarantee of the personality of any one dog – nature and nurture are inextricably intertwined. But just as the Siberian silver foxes were selectively bred for tameness, different breeds of pedigree dogs have been bred for particular traits through generations and these often still pop out in their modern descendants. Our girl, Ripley, is a second-generation Cockapoo and we'd have to clamber high through the branches of the family tree to find any ancestor that might have been bred to point out game birds. I confess I know little of the world of hunting and shooting and yet I think I can detect some hints of her roots. She's busy, energetic and people-focused. If we're distracted from our essential job of picking up the ball on the beach for a throw and it rolls away down the shoreline, she'll do a lovely point: elbow crooked to signal where it's gone. If we still don't snap to it, she'll add an irritated nod in the general direction: 'It's there, you idiot!' This nod, I suspect, is not a trait inherited from any hunt-accompanying ancestor, but something entirely her own, the result of living with two humans who fail to focus fully on the priorities in life.

While the likelihood is that you don't need a dog to protect you on your rounds collecting taxes (Dobermann), haul your fishing nets back to the shore (Newfoundland), hunt wolves (Borzoi), flush out badgers (Dachshund), fascinate waterfowl (Nova Scotia Duck Tolling Retriever) or perform tricks in a travelling circus (Miniature Poodle), knowing something about a breed's 'jobs' and traits can give an indication of how easily a particular type of dog might fit into your lifestyle.

Pedigree breeds tend to be organised into groups based around their working 'purpose'. There are the sporting dogs: pointers, Weimaraners, Vizslas, spaniels, setters and retrievers. These were originally bred variously to point out the game birds, flush them so they fly up from cover and retrieve the downed birds for the hunter. They tend to be active and energetic, lovers of water and woodland. And because they were bred to accompany people and other dogs on the shoot, they were selected for sociability and people focus. Some of these traits may have become muted over generations, but it's likely that energy, enthusiasm and sociability will be part of the mix. And a love of getting good and muddy.

Though again, in all of this, nature and nurture will both play a part; if they haven't had that all-important early socialisation then sociability may be something of a lottery.

GREAT EXPECTATIONS

Still on the hunting trail, are the hounds. Scent hounds, like beagles, bassets and fox hounds, were valued for their ability to pick up the scent of game and follow the trail. Sight hounds, like greyhounds, whippets, Salukis and wolfhounds, have their eyes, rather than their nose, on the prize and the speed to pursue. Hounds, when they've picked up the scent or sight of something worth chasing, are likely to be laser-focused on the 'prey' and hard to recall or distract, which means hound owners need to have their wits about them. Scent hounds have stamina rather than speed and most enjoy a good long walk. Sight hounds are sprinters, who, though born to run very, very fast indeed for short periods, are then more than happy to have a nice, long sleep, Greyhounds are reported as being inclined to snooze for eighteen hours a day, which makes them pretty competitive in the sleep stakes with the average cat. Sadly, given their high prey drive, without an awful lot of training and vigilance, cats won't find them the best of bedfellows. And don't be totally taken in by their reputation as dedicated couch potatoes. Having given occasional training support to a greyhound rescue, the dogs enjoyed stimulation and would happily spend a couple of hours, running, playing and training in the paddock.

Then there are the herders or 'pastoral' breeds, who, as the name suggests, have long been with us, moving livestock around and, in some cases, protecting them, too. Collies are the obvious candidate that comes to mind. Intelligent and hugely energetic, they need bags of exercise and hate being bored. If you've read one of those stories about a dog that has learned a record-breaking vocabulary, it will almost certainly be a Collie. But even with these skills, clearly you can't leave them to their own devices to get stuck into a good book – they need plenty of mental stimulation to keep them happy. The German Shepherd and its Belgian cousin come into this group too. Because they were bred to guard the flock, they can be 'spooky' and wary of strangers, as the Bidens and their security detail found to their cost.

Then we've got the terriers, originally relied on for pest control: hunting and often digging out 'vermin'. There is a terrier variety to suit any size requirement, from the small (Jack Russells, Westies and Borders) to the more sizeable English Bull Terriers and Staffies, Soft-coated Wheatens, Lakelands and – largest of all – Airedales, who were used as police dogs before the First World War and as fearless messengers in the trenches during it. And that's only a small flavour of all the different terrier breeds. There are some that need little or no grooming to the long, curly or double-coated breeds that need lots of dedicated coat care. It's such a huge group of dogs that trying to paint any patterns of personality can certainly only be with the broadest of brush strokes. Because they were originally bred to hunt vermin, they've got lots of drive and a powerful chase instinct. But they can also be playful, affectionate and mischievous too.

If the terrier groups are a fun-loving bundle of diversity in size and personality, the utility group are even more so, encompassing everything from Miniature Schnauzers and Boston Terriers to Dalmatians and Bulldogs. All were, in theory, developed as a breed for a purpose so they're bundled all together, but with very little in common.

The 'working' dogs group is again a catch-all, covering breeds who did service hauling sleds, rescuing humans lost in the mountains or guarding. Included in this camp are the Huskies, St Bernards, Dobermanns, Boxers and Rottweilers. When it comes to the latter, I am, of course, deeply prejudiced myself but I have to say that our Rottweiler, Murphy, never exactly struck me as the 'working' type. He was definitely more of a grifter – sharp as a tack and fast on the uptake in training, but always keen to deploy what he learned to his own advantage; he would learn all day if you made it worth his while. Nevertheless, when faced with a situation where he felt members of his family were under threat, he could rock back on his rear legs, bounce and bark. In the extreme, his guarding instincts hadn't disappeared.

The toy breeds weren't expected to work for their living. They were always intended to be companions and, sometimes, professional lap warmers or alert dogs (members of the Chinese imperial court were said to hide miniature Pekingese in the sleeves of their robes for just this purpose). In the toy group are the Bichons, Havanese, Maltese, Chihuahuas, King Charles Spaniels, Italian Greyhounds, Pugs and Yorkshire Terriers (who, yes, are illogically called terriers but classed as a toy breed). Most – though by no means all – toy breeds don't require much exercise, but that doesn't mean they don't need attention and plenty of playtime. Unsurprisingly, they like to be close to their human, having been keeping people company for centuries. You'll see them peeking out from behind or nestling close to the skirts of noblewomen in Renaissance art and some have quite the impressive contact list in their ancestral black books. Miniature or Italian Greyhounds were beloved of Catherine the Great and Queen Victoria. Madame de Pompadour favoured the Bolognese. And, of course, King Charles II did love a miniature spaniel. He took his everywhere, which caused diarist Samuel Pepys to comment disapprovingly on his silliness in playing with his dog at meetings when he should have been paying attention to the business of state. I, however, would never condemn dog-related silliness, royal or otherwise.

This canter through breed purposes can only take us so far. Each dog is his own canine, with a genealogy all his own that drives his temperament. It's just a starting point and, of course, the original purpose of the breed is not the whole of what determines behaviour.

We're learning more and more about the influence of genetic make-up. Labradors have long had a reputation for being greedy eaters and, for some at least, it really is down to their genes. A recent University of Cambridge study[1] found that 25 per cent of Labradors and a whopping 66 per cent of Flat-coated Retrievers have a genetic mutation that means that they feel hungrier between meals than other dogs, but at the same time burn 25 per cent less energy when they're at rest. This means they're constantly craving food and it's all going straight into that spare tyre round the middle. That is a cruel trick of genetic fate. I know how they feel.

Sadly, when dogs were selectively bred for skills and personality traits or particular looks, health problems came along for the ride. Hip or elbow dysplasia is a risk for many breeds. King Charles Spaniels and Griffon Bruxellois can inherit the painful condition syringomyelia, where the skull isn't large enough to accommodate the brain. Inherited eye, hearing and heart diseases can be an issue for some breeds, too. Flat-faced or brachycephalic dogs, such as Pugs and French Bulldogs, have been so extremely bred over time for a shortened skull that it has brought with it any number of health problems, including breathing difficulties.

Screening programmes are in place for many of the conditions to try to ensure that only healthy dogs are allowed to breed, which is why a pedigree puppy's details will, or should, give details on screening tests carried out on parents, for example, the canine parents' hip or elbow scores.

Nevertheless, problems do continue. Concerns about the health issues of Cavalier King Charles Spaniels have even led to a ban on 'pure breeding' in Norway, who now encourage only science-based cross-breeding programmes. Elsewhere, 'retro pugs' are being introduced: pugs crossed with Jack Russells to lengthen the snout so that it's closer to the type of dog you would have seen in the nineteenth century. The health issues of selective breeding is undoubtedly one of the worst examples of unhelpful human 'nurture'. Whether driven by passion or profit, meddling too much with nature can't be good for our dogs.

THE RISE OF DOODLEMANIA

Beyond the pure pedigrees, there has been an explosion of poodle crossbreeds in recent years. These are often referred to, sometimes with, perhaps, just a hint of a sneer, as 'designer dogs'. I find this puzzling as purebred pedigrees are all essentially the result of breeding by design. The 'creation' of the Labradoodle is credited to Wally Conron, who worked for the Royal Guide Dogs Association of Australia in the 1980s and was asked for a guide dog by a blind woman in Hawaii whose husband had allergies. But there were doodles around way before that. The author Monica Dickens bred a Goldendoodle (Golden Retriever-Poodle cross) back in 1969 and, without doubt, there were others out of the public eye doing the same. Still, it's probably fair to say that the doodle variants' popularity started to take hold when Conron's Labradoodle litter was actively promoted as hypoallergenic.

Conron later said creating the Labradoodle was the biggest regret of his life. Their popularity had grown so wildly, he mourned, that it had spawned an unscrupulous backyard industry, breeding puppies with behavioural and health problems. But this isn't something confined to doodles. Unscrupulous breeders have long followed hard on the heels of popular enthusiasm for particular types of dog.

Yet despite Conron's very public breeder's remorse, popular they have continued to be. Since then there have been poodles crossed with all kinds of different breeds: Cocker Spaniels, Cavaliers and Golden Retrievers, of course, as well as Schnauzers, Jack Russells, Bernese Mountain Dogs and Malteses. In online pet website Pets4Homes' most popular UK dogs listing of 2023, four poodle-crossbreeds made the top 10. Some poodle crosses are low shedding, which means they are indeed less likely to cause allergies, but types of coat can vary considerably. The silky, wavy coats may still shed; woolly or fleecy-coated dogs much less so. That doesn't mean their coats are hassle-free. Keeping their 'do' in good condition can be like trying to brush and style a sheep. They need serious salon time.

In personality, with all the usual caveats of nature and nurture, socialisation and training, poodle crosses are playful, affectionate, fun-loving and healthy dogs. It seems tragic to me that Wally wishes he hadn't brought Labradoodles and their crossing cousins into the world. They've brought joy to an awful lot of people, including me.

Then there are the mixed breeds in all their wonderful diversity. Often you can see hints of their parentage – or grandparentage – in the shape of the ears, the length of legs or set of the tail, but these are dogs that defy grouping and classification. Because they haven't been purposefully bred, they can swerve some of the genetic disorders that affect specific breeds.[2] Their temperament will still be the product of nature and nurture. If you know the parents and the environment in which they've been raised, you can tell much from that, but there's no easy accessible crib sheet for characteristics. Whether you call them mutts, mongrels, mixed or beautifully blended, they're none the worse for that.

THE PUP INDUSTRY

Whatever the breed, finding the right dog calls for some care. Getting a new dog should be a time of excitement and anticipation and, for most, it absolutely will be. But there is a need for healthy caution – increasingly so. The pup industry is big business and, in its darker corners, as lucrative as it is ugly.

When it comes to finding a dog, whether a puppy or an older dog, most of us initially hook up online. According to the PDSA's 2023 PAW Report, 72 per cent of owners found their new dog online via pet websites, rescue websites, breeders' own sites or online advertising sites – not that surprising given we're in a digital age. But as with all long-term relationships, that initial online contact should be just the beginning of deciding whether this is the right dog. You need to meet in person.

The pitfalls are most acute with puppies. Illegal puppy farming has always been a money maker, but since the pandemic has become even more so. The scale of it is largely unquantifiable because it operates in the shadows, but one or two animal welfare organisations have had a shot at it assessing it. The Scottish Society for the Prevention of Cruelty to Animals in 2022 estimated that illegal puppy farming was worth £13 million in Scotland alone. Meanwhile, Naturewatch Foundation believes 400,000 illegally farmed puppies are bought every year. Aside from home-grown puppy mills, dogs are also illegally trafficked from abroad. Puppies bred in this way can have inherited health problems and they will almost certainly be under-socialised and not habituated to the normal environmental sounds they will encounter later in life, making behavioural problems more likely.

Recent legislation in the UK has tried to make it harder for puppy farmers and illegal importers to carry out their trade. The law requires anyone breeding three litters or more a year or breeding for profit to be licensed and has introduced new safeguards around sales. Lucy's Law (named after a Cavalier King Charles Spaniel who was rescued from a farm after being kept in appalling conditions, breeding litter after litter) came into force in 2020 in England, followed by Wales, Scotland and Northern Ireland; it bans the sale of puppies through third-party dealers or pet shops. Legally, puppies can only be sold in the UK directly by the breeder from the location in which the puppies were born and reared (rescue organisations are exempted) and potential new owners must be allowed to see the mother and littermates. It hasn't completely stopped the practice – criminal gangs will always find a workaround. One of the latest tactics, as people have become more aware of puppy farming, is to use a network of associates as fronts to skirt the breeding legislation, placing puppies in what appear to be ordinary 'family households' with dogs that aren't their mother so that they can be viewed for sale.

Organisations seeking to combat illegal puppy breeding, including the RSPCA, Dogs Trust and British Veterinary Association, have taken further steps to safeguard standards by producing a 'puppy contract' for breeders to offer to prospective parents, which sets out details of vaccination, the puppy's parents and health screening tests and, equally importantly, where the puppy has been kept and the environmental experiences to which the pup has been exposed, e.g. people, other dogs, household noises. This provides a new owner with a solid foundation on which to build in those important early weeks and months. Obviously, even this doesn't act as a guarantee, but it is a good indicator that this should be a responsible breeder.

Welfare and breeding organisations are also working to make the advertising of dogs for sale more responsible. The guidelines of the Pet Advertising Advisory Group require advertisers to label the origin of adverts so that it's clear if they are private sales, licensed breeders or rescue organisations. They also expect advertising sites to monitor private sales listings for multiple repeated phone numbers and email addresses to weed out puppy farmers posing as private individuals. The Kennel Club Assured Breeding Scheme certifies that breeders have been inspected for welfare standards and carried out the relevant health screening tests on the dog parents.

Will all this mean illegal and unethical puppy breeding goes away? Sadly, the answer is no, it won't, but everything that makes it more difficult makes it less profitable and less attractive as a way of making money. That is the best we can hope for.

The dog that holds the key to your heart, of course, does not have to be a puppy. More than 100,000 dogs are taken into rescue organisations in the UK every year. Navigating the world of the rescue deserves another chapter.

CHAPTER 13

THE RESCUERS

Starting out a relationship with a second-time-around dog is, by anyone's standards, a good thing to do. It's giving a dog that would otherwise be living out their days in a shelter (or worse) the chance at a life of love, fun, mischief and adventure. But however fine and principled a thing rehoming a dog is, I'm not about to be evangelical here. It still has to be the right decision for you. For some, getting a particular breed of puppy, rather than a rescue will be the right choice. Let's face it, when you get a dog, you're looking for a long-term chum who will be hanging out with you, meeting your friends and family, blithely eating their way through your disposable income and demanding your attention – yes, even at that really tense moment in the TV thriller or football match. If you aren't confident about being able to love and commit wholeheartedly to them, feeling righteous isn't going to be a substitute – for you or the dog. Better to make a donation to a rescue charity, instead.

The good news is that for thousands of people and dogs every year, second-chance love is exactly what they need in their life. Rescue dogs are usually toilet trained so no having to get up in the middle of the night to wander around the garden in the cold and the rain, just praying for a pup to perform before you get hypothermia. They may have had a little, or even a lot, of obedience training with a previous owner or at a rescue centre or foster home. They'll be vaccinated and health-checked. You might be offered an idea of what they love or hate: other dogs, cats, playing ball or tug, going for a run or just pottering around a park. And most will come with a character reference, so you know something about their personality and what you're committing to – what's not to love?

Some rescue dogs, like my own Thomson, will have had bleak pasts that have formed their attitudes to what they encounter in their present. They are by no means the majority of dogs looking for a new home, but these dogs need special care. Yes, it can be hugely rewarding, but resilience and infinite patience, dogged determination and the ability to cope with the ups and downs along the way need to be part of the human profile for these more damaged souls.

What you shouldn't demand or expect with any rescue dog is that they will be automatically 'grateful' or 'eager to please' because you've given them a better life. It's up to you to make the bond. It doesn't come for free.

These days, it's estimated that in the UK alone there are more than a thousand organisations dedicated to rescuing and rehoming animals, ranging from the large, national charities to passion-project independents, the take-all-comers-generalists to the breed-specific, all vying for a share of dog lover attention.

It wasn't always so. Until the early nineteenth century, life for dogs could be nasty, brutish and short. They were largely kept to work and most lived on scraps or on their wits. The wealthy and the nobility might have pampered favourites, but, for pretty much everyone else, keeping a dog as a pet was considered frivolous, a luxury they literally couldn't afford.

But attitudes towards animal welfare started to change. Keeping pets became more socially acceptable and was even viewed as morally uplifting. People began to be more emotionally invested in their animals and dogs began to move out of the kennel and the yard and into the home. With this shift in attitudes, came the rescuers.

The earliest animal welfare charity in the UK was what would later become the RSPCA. Founded in 1824 by a London vicar, Reverend Arthur Broome, and a coffee house group of gentlemen that included anti-slavery campaigner William Wilberforce, it was not, either then or now, centred on dogs, but focused on 'the suppression and prevention of wanton cruelty to animals'. A large part of the initial activity was focused on conditions in Smithfield meat market in London. The group's initial efforts did not go so well. Reverend Broome personally acted as guarantor for the debts of the Society and as these grew beyond his means to settle them, he ended up with a spell in a debtor's prison.

Attitudes to animal cruelty were slowly changing, but not being able to pay your creditors was still way beyond the pale. Yet within a few years the Society was back on its feet, had Princess Victoria as patron and would, by 1835, be granted its 'royal' status.

The activities of the animal rescuers gathered pace through the Victorian age. The first dogs' shelter was established in 1861 and would become Battersea Cats and Dogs Home, followed by the RSPCA's first shelter in Liverpool in 1883, the National Canine Defence League (now the Dogs Trust) in 1891 and Blue Cross in 1897.

Those Victorian 'start-ups' are still the biggest hitters in the UK dog rescue world, but they've been joined by hundreds and hundreds of others. Surprisingly, anyone can start a rescue shelter. Unlike commercial boarding kennels, there's no requirement to have a licence and there are no regulations. Not all operate rescue centres. Some, often breed-specific rescues, facilitate direct matches between potential adopters and people who need to rehome their dog, which means there can be opportunity to speak to owners at first hand to learn about a dog's personality and behaviour. Others work with a network of fosterers, so dogs stay in a family home until their forever human comes along. These fosterers are, to my mind, the heroes of the dog rescue world. When, occasionally, they fall in love with a dog in their care and decide to adopt, they describe it, wryly, as a 'foster fail'. Strange – it seems like a spectacular success all round to me.

MAKING A MATCH

Finding a new dog soulmate can take some dedicated research and not a little admin. Dog rescue organisations set their own procedures for adoptions and they can differ considerably. They reserve the right to decide who's the right fit for a particular dog and often don't publish all the criteria they use to assess prospective adopters. This may be to avoid prospective owners 'gaming' the system or because, understandably, they don't want to be drawn into time-consuming arguments or challenges. They do have first-hand knowledge of the dog, after all. But not knowing all the 'rules' can, without doubt, lead to frustration, anger or a sense of rejection if people are turned down when they feel they could give a dog a good home and aren't sure exactly why the rescue appears to disagree. In general, the main criteria tend to be around the age of children in the household or as regular visitors, other resident pets and how long dogs will be left alone, as well as living arrangements and the previous dog experience of adopters. There are some welcome indications that criteria around using dog walkers and dog day care has been updated to get in touch with modern life, allowing people with jobs to adopt. Round of applause for Blue Cross, particularly, for being loud and proud about that.

The initial form-filling, often completed long before getting within sniffing distance of a dog, can feel a little reminiscent of the palm-sweating angst of job or university applications. There's even sometimes a space for what seems very much like the 'personal statement' beloved of UCAS forms, setting out what the applicant can bring to the life of the dog.

But guess what? In some ways, a 'job' application is exactly what this is. And as with any good job application process, it should be a two-way thing. Prospective adopters, like smart job applicants, should be buzzing with questions, probing any details of the personality description that aren't entirely clear so everyone shares the same understanding. 'Great when you get to know her little quirks.' What are the quirks, exactly? 'Takes some time to settle with new people, but then he'll be your best friend.' How does that manifest itself – hiding away or growling? 'A hands-off approach is best to win his trust.' What level of handling can be tolerated – any or none? Is it possible to get a collar and lead on? Some rescues provide gold standard, transparent descriptions of dogs and the kind of home they need, others are more creative and open to interpretation. It's in no one's interests if a dog is returned so questions should be welcomed.

Some of the larger charities, like Dogs Trust and Battersea no longer accept applications for a specific dog, instead inviting prospective adopters to indicate the kind of dog they're looking for but taking on the whole matchmaker role themselves. Battersea even automates the initial screening of applicants. This feels, sadly, like they're borrowing from the more chilly and soulless practices of the human recruitment market, but at least they're upfront about it. Independent rescues set their own criteria, based on their experience and perspectives of humans and dogs. They will have strong opinions of the needs of the dogs they're seeking to rehome and sometimes strong personalities to match, but the great ones can be extraordinarily generous with information and support.

Once the paperwork has been completed, generally the process includes a 'home check' of one kind or another to take a view of the living accommodation and eye up everyone in the household in their natural setting. It seems here, particularly, that an awful lot of anger and controversy is stoked up, particularly around blanket requirements for six-foot-high fences to gardens or proximity to busy roads. There is some sense to this, as in the first weeks a dog may feel unsettled and uncertain and try and bolt, but it can seem incomprehensible to experienced past owners whose four-foot-high fences have stood the test of generations of dogs. In-person home checks are often carried out by volunteers who may not provide checks exclusively for one rescue, so they are likely to have to abide by the framework given and sometimes have relatively little scope for reasoned flexibility.

Despite the administrative hurdles to be vaulted by the human side of the partnership, one in five dogs was adopted rather than bought in the UK and set off from shelters to a new life and home.

MORE THAN JUST A DOG

CROSSING CONTINENTS

Of course, apart from the national rehoming charities, there are also the organisations who rescue dogs from overseas – street dogs, stray dogs, abandoned dogs and hunting dogs no longer regarded as useful for working. Overseas rescue has become a subject that has given rise to high emotions and angry debate in recent times, as I found when I got involved with Sophie from Romania, adopted by former BBC technology correspondent, Rory Cellan-Jones. Fiery rescue advocates take up battle lines against equally fervent vets who believe importing street dogs risks bringing diseases into the country that aren't found in the dog population here. Both camps are fully prepared to turn their fire and their ire on trainers like me, because in this 'if you're not for me, you're against me' world, I offend both sides.

I'm quite prepared to acknowledge that I'm not qualified to comment on the scale of the health risks; I'm not a vet, let alone Chief Veterinary Officer. But what I do know about is dog behaviour. Dogs who have never been exposed to the kind of environmental stimuli they are likely to meet in a UK home may well have some behaviour issues that will need patient and gentle handling in the early weeks and even months. (The PDSA Animal Welfare Report 2023 found that 31 per cent of owners of dogs rescued from abroad reported fear-related issues, compared with 16 per cent of dogs from UK rescues.) Denying that those early behaviour problems might exist is irresponsible nonsense. But, equally, raging that dogs already adopted shouldn't be here doesn't help the people who are trying to cope with a dog who is in their home with their family and who are finding it a struggle and need support. If the dogs are here and here legally with the approved health checks and vaccinations, I'm willing to help advise on behavioural issues if and when I can because they can usually be resolved, or you can at least improve the situation significantly.

The largest percentage of imported rescue dogs are from Romania, according to a University of Liverpool study, followed by Cyprus and Spain.[1] There are reputable rescues rehoming dogs from abroad who follow the advice of the Government's Chief Vet and carry out comprehensive blood tests before the dog leaves the home country and again before rehoming in the UK. Since Autumn 2022, rescue organisations from particular 'high-risk' countries (which as of 2024 included Romania, Poland and Belarus) have to apply for Approved Importer status and abide by stringent conditions on certification of health checks carried out. To keep all our dogs safe, however heart-rending the story of need that's presented, this seems like a minimum standard that anyone considering adopting from these countries should demand and it is more than reasonable to ask to see proof of compliance.

Some overseas rescue dogs settle into their new home pretty easily. For others, particularly those who are fearful of people, it can be difficult to help them see their new home as a safe base when the first time they encounter their adopters is when the van pulls up outside the front door. They can absolutely make progress, but it will take more time and patience. It's worth saying that just because a dog is from abroad doesn't mean they learn in a different way from any other dog. Their early experiences will have undoubtedly helped forge their behaviour. This might include separation from their mother when too young, poor early socialisation, limited habituation to different environmental stimuli (traffic noises, building work, vacuum cleaners, even the clatter of pots and pans in the kitchen). They might also have been badly treated if they've been living on the streets, but by no means all dogs with behavioural issues will have been abused. Whatever has formed their behaviour, they will still learn to overcome fears in the same way as any other dog: through classical conditioning (showing that something they fear predicts something positive) and they will still learn to rock their obedience behaviours through operant conditioning (if they do this, and not that, they earn a reward). There's also no need to worry about learning another language. Dogs follow visual cues first (e.g. hand gestures) and you add a verbal cue only when the behaviour is well dug in. It's for convenience more than anything else; at that point, you can just match the behaviour with any word you want.

THE OLDIES

Often overlooked in rescue are the dogs described as 'seniors' in their matchmaker profiles. Dogs tend to be classed as 'senior' from about the age of eight, which seems harsh, coming from a species that endlessly pushes back the age at which any of us are prepared to be labelled 'old'. For many breeds, eight is really just middle-aged and many of these dogs still have plenty of active fun to enjoy before they're at the pipe and slippers stage. Even the genuine oldies can be very much young in spirit, if slightly slower in body. Sure, they'll have formed their personalities and know what they like and what they think they don't, but they'll still be up for forming a bond if you open up your home and heart. And, yes, they very much can learn new tricks.

Of course, there is no getting past the fact that adopting an older dog – whether a midlifer or genuine OAD – means accepting that you'll have less time to enjoy their company, though maybe not as little as you fear. The indications from recent (albeit limited) research are that our dogs are living longer and staying healthier, thanks to improved nutrition and veterinary care.[2] But though the oldie adopter may have more good years with their dog than they might have done in the past, they will face that heart-wrenching moment of goodbye all too soon. I've tried to think of something, if not upbeat, at least encouraging to say about this, but, in truth, I can't. It will hurt. I can only lean into the eternal truth that grief is the price we pay for love and the love for oldies, while it might not burn long, can burn very bright.

IS IT LOVE?

Whether the dog is homegrown or rescued from abroad, my feeling is that it's a brave man or woman who will take on a life companion without ever having met them. The big national charities insist on the whole family (including any resident dogs) meeting their potential new family member. Good independent rescues, including some who bring in dogs from abroad, will also encourage a meeting either at their centre, if they have one, or at a foster's home with arrangements made to pick up the dog on another occasion if all goes well. This seems to me to make absolute sense. Others operate what I'd describe, perhaps unfairly, as 'click and collect'. Applicants find a dog on the website, are screened, home-checked and then invited to come and take the dog home. This causes me some discomfort as it puts an awful lot of pressure on the adopter to go through with the deal, even if they find the dog is not a good fit when they meet them. Other charities offer descriptions, photos and videos for prospective adopters to make a decision and then make arrangements for the dog to be transported to the new family, which means the first time they meet is when they're handed over on the doorstep. Again, this has potential for stress all round at the beginning, so it's absolutely crucial to factor in a period of adjustment while everyone gets to know each other.

Reputable charities will keep in touch with adopters to make sure their new dog is settling in and some, including the large national charities, will offer behavioural support, if needed. This is important and adopters shouldn't be shy at taking up the offer. Early intervention can dial down the anxiety levels, put problems in context and help resolve issues that can get in the way of building a strong bond. Good rescues will also allow adopters to return the dog if it really isn't working out. While essentially 'click and collect' and 'drop off' adoptions would personally give me reason to pause, I'm not in the business of finger-wagging or, at least, I try to resist the temptation. All I would say is that it's smart for any potential adopter to read the signs and consider if there are too many red flags with any rescue. Are you able to meet the dog before committing? How and when will the meet take place? If you can't, what's the quality of the information you're given? Are you and your questions treated with respect? Will you get any support and will the rescue take back the dog if it all goes wrong? A good rescue will vet you quite meticulously to ensure that you are the best fit for the dog so there's no harm in returning the compliment. The rescue is looking for the right home for the dog. You're looking for the right dog for you. It's in everyone's interests to get this new relationship off to the right start. And if you have the opportunity to meet the dog several times before taking them home, that's all to the good.

And when the newcomer arrives at the home, what to expect? Even if you've met the dog at a rescue centre or foster home, for dogs who navigate and understand their world through their highly developed senses of smell and hearing, there will be an awful lot about your place that is very, very unfamiliar. The routines will be different, too. You can't just sit them down and have a chat about what they like doing in their spare time over a cup of tea or a glass of wine. (Though you may very well be gasping for, and completely entitled to the tea or wine.) You just have to let them navigate the newness and help them to recognise you, your people and your place as their safe base. It's about taking a deep breath, holding your nerve and asking for help if you need it. The adjustment will take place in its own good time. You can't hurry love.

THE RESCUER'S STORY

I've been providing behavioural and training support for Black Retriever X Rescue in Wiltshire for almost ten years now. It was founded by campaigner and documentary film-maker Jemima Harrison, who produced the BBC *Pedigree Dogs Exposed* documentary on the health and welfare problems of breed standards. She set up BRX more than twenty years ago. This is her take on running an independent rescue.

'I started out literally with the idea 'I love dogs and I think I understand them'. I've always had a good instinct about dogs, but what I knew then was nothing compared with what I know now. I read a lot around dogs and there is so much new knowledge out there. In fact, one of the things I always tell would-be adopters is that getting a new dog is a great time to refresh your training and behaviour knowledge.

'Running a rescue is all to do with loving dogs with an obscene passion and taking that obscene passion, that might otherwise leave you lying in an emotional ditch because you care too much, and using it for something practical – something that can properly help them. I want to understand each dog as an individual and find the best home I can for them. I write up their stories and how they've come to the rescue in a way that's as honest as possible. It gets emotional buy-in, of course, but it's also about transparency, which is hugely important. After they've been rehomed, I follow through with updates on Facebook about how they're getting on. It acts as a learning exercise for people of what can be involved.

'Rescues get a lot of flak for box-ticking and not being flexible when it comes to finding the right home, but at the end of the day, the dog comes first. We ask prospective adopters a lot of very specific questions to get a really rounded view of the life that dog will be going into. We try to make it easy for people to fill in the Q&A and then we'll talk to them on the phone. You get pretty expert at reading the responses and we have to be – for a nice young Labrador, we might get more than 200 applications so we have to sift quite quickly. Then these days we tend to use home-check videos. I give people thorough instructions on how to take the video and you can learn an awful lot. If I still need more information, I can send someone in person to follow up. But by that stage, we tend to have built a strong relationship with the people who we're looking at to rehome a dog. We'll have had lots of exchanges.

'I want to find the right home, but one thing I've learned over the years is that not every home has to be absolutely perfect. You can turn a home that isn't into a perfect home with the right support. I use WhatsApp a lot because people know they can get a quick response from someone who knows what they're talking about. We give adopters a guide to settling in, which basically tells people: "Don't panic. There will be times when the dog will drive you nuts, but things will settle down. Call us, call us, call us if you need help. We aim to tailor-make support so that it's specific to that dog."

'A huge thing for me is not to be judgey about people surrendering their dogs. It serves no purpose whatsoever. If we can, we keep in touch with the original owners to tell them how their dog is getting on and send them pictures and videos. We've got a good reputation as a result. People recommend us because they say: "They were nice and didn't judge us and they're still sending us updates five years on."

'Not everyone is willing to take a dog with obvious behavioural challenges. You have to be careful not to overcommit with the dogs you take for rehoming. You might have ten easy dogs to one tricky one, but you don't know which one will turn out to be tricky until they're in your care. Most people just want a really nice, brilliantly assessed dog. But there are some people who genuinely want to learn and to help. We had a brilliant case of a couple who contacted me saying that they were in a position in life where they wanted to take on a dog with behaviour issues and they thought we were the rescue that could support them to do that. They had no real behaviour experience. I've just had a message from them with a photo, saying they'd just taken Bear for his first pub lunch.

'They've adored the whole process, but people like that don't come along very often.

'Advice for potential adopters? Don't sweat the small stuff. An awful lot resolves itself. Let the small indiscretions go. Dogs are like chameleons. They've been honed over millennia to inveigle their way in with us. They're really fantastic at picking things up. Don't panic because you're on the sofa with the new dog and he growls at your old dog. There's every chance it will have resolved within a week. But don't struggle on your own, ask for help if you need it. And as soon as you can, get them out for a bloody good run. For every single rehome, we check out the local secure fields and let the adopters know where they are.

'I wear other hats. I'm a campaigner. I'm a film-maker. I'm bringing all kinds of experience from outside into making this a better rescue. I wish there were forty-eight hours in the day. But there is nothing that gives me the same fulfilment as running the rescue. Making films is a great passion, but there are days when I don't want to get out from under the duvet because I know it's a particularly tough day ahead or there's a tough problem to be solved. I've never ever felt that with the rescue. I can get to 10 o'clock at night and think: "I can park that until the morning." And the next day, I'll be ready to start again.'

PANDA'S STORY

The van rocked up about mid-afternoon as expected and parked a short distance from the house. It was well appointed, properly ventilated and designed for four dogs in total, I think. This wasn't mass transportation, and I soon got a sense of how committed to his job and how skilled the driver was. He clearly described what I should expect of his passenger, and we formulated a strategy before he opened up the van. We were clear what each of us would do and where we were going. The journey had been relatively short but that had made no difference to our new resident to the cabin. She emerged like a Tasmanian Devil, lurching this way and that. When she saw me, the spinning started too, along with the unmistakable barking of an alarmed dog.

The cabin was set up exactly for this kind of arrival. There were 'air-locks' everywhere that ensured that the resident couldn't escape and allowed us to isolate them in various parts of the set-up should we need to. Between us, the driver and I got Panda into the run that ran the length of the cabin side. The gate locked, I said a heartfelt thank you to the driver and turned to look at the new arrival.

She was Panda by name and appearance, but not so much by nature. There was nothing of the cheerful bumbling of the bears trying to outwit their keepers that pop up on TikTok videos. As for her breed, there was unquestionably some Newfoundland in her and perhaps some Collie, but beyond that I'd have to give up on the 'guess the breed mix' game. I'm almost always surprised.

At about a half to three-quarters of the size of a purebred Newfie, Panda was thick-coated and strong. I was later to describe the first walk with her on lead as like handling twenty ferrets in a force eight gale. What can I tell you? What a treat!

I could stand and look at Panda from about three feet from the run enclosure, but if I moved even a little, she exploded into a rapid-fire bark and would throw herself against the enclosure panel. If I dropped chicken from above her through the bars of the enclosure, she would retrieve it and then resume the real business of telling me just what she thought of me. I had built the run well and I was, at that moment, glad of it.

I'll just say a little about how Panda came to be with us. Panda had been relinquished to a rescue that was based close to us and run by the formidable Jemima Harrison, at that time a film documentary maker, as well as a campaigner for animal welfare and the operator of a shelter that usually had around ten free roaming dogs in residence. Jemima felt that Panda was going to be a challenge to integrate easily with the other dogs in her care until she had been fully evaluated and some behaviour modification was put in place. And so she came to be with us.

She had been relinquished for two pretty significant reasons. She wasn't great with strangers, but there was also a slightly confused account about her picking up and shaking her canine companion in the house. The details were vague, but an assault on a smaller dog had to be taken seriously. As far as we knew, the smaller dog had not sustained any obvious injuries in the encounter, and it was not altogether clear whether Panda had actually picked up the other dog or not. All the same, we had to proceed on the basis that she had. This might, ultimately, mean no access to small dogs in her future but didn't necessarily rule out dogs altogether if she proved to be prosocial.

There are a lot of things that I could say about the force of nature that is Jemima. One of those is that she has never shied from taking on challenging dogs; another is that she places a great deal of importance on understanding, evaluating and modifying behaviour in the dogs in her care. After all, the most common reason for dogs being returned to shelters and rescues is behaviour. I'm not the only trainer who supports Black Retriever X Rescue and who can support adopters whenever they need help. It leads to more adoptions succeeding and more fosters turning into adoptions. It serves as testament to Jemima and her rescue that she manages to successfully place as many challenging dogs as she does.

But I digress. Back to the problem in hand. The whirling dog currently standing in the run was giving me the side eye. Slow movement was followed by an explosive response, which was followed by chicken drop, followed by calm. Rinse and repeat. Of course, the scenario wasn't ideal. Panda was clearly over her fear threshold. I could not get far enough away from her to get her under it. I could throw the food over the top of the run panelling, but that made little difference. Nevertheless, I made progress that first afternoon. Eventually, I was able to gently feed her through the bars and she would even offer a default sit to tempt me to up the pace of chicken delivery.

This was as near as possible to a standard desensitisation and counterconditioning protocol as I could get, with me working my way closer to the enclosure until Panda was eating from my hand. But while I could clearly make progress in creating positive associations with me, thanks to the delivery of chicken (Panda had some very pronounced indicators that she had formed a positive conditioned emotional response to each step), if I went out of sight and reappeared it was as if we had never met. We were back to maximum distance and throwing the chicken as near to her as I could manage.

Settling Panda down for the first few days meant fashioning a 'pig board' from some plywood and manoeuvring her through whichever door was most accessible, having refreshed water and food as necessary. Now this was high-end accommodation, with a two-seater sofa, storage space, a comfy chair, kitchen area, TV and a dining table with four chairs. Don't laugh. There were always going to be times when simulating a home environment was going to be important and, besides, why shouldn't they all feel like they are in a comfortable place? But as far as Panda was concerned, this was still alien territory.

Every time I started a new session with Panda we would have to go back to the beginning of the plan. It was as if I were a stranger again. We would make progress in the same way that we did in previous sessions, but no further. The conclusion, ultimately, was that Panda suffered from a relatively rare condition called Sudden Environmental Contrast. There's often a fear of strangers component to this condition, but as conditions they tend to be second or third cousins, rather than sisters. I won't go into depth about how SEC presents, but it's broadly to do with things that change suddenly and, for the dog, unexpectedly. It might be someone moving from stationary, or an object that wasn't there yesterday, or someone or something behaving in an odd way. Despite being very much in the 'upset' camp, SEC doesn't respond to the standardised plans for fear. You have to 'break the rules' and accept that the dog is going to go off each time you approach and carry on until the dog is calm and taking food without going off, rest them and then start again. In every session, you're looking to reduce the time it takes until calm is established. That's too brief an explanation of how SEC training works, but it will need to suffice for now. While it does break the rules of desensitisation and counterconditioning, the unfortunate reality is that nothing else works. Equally important is that not all dogs with this condition respond to any training at all, which may mean a lifetime of careful management.

We were fortunate. Panda responded very well and continued to make progress. I became the best of chums and, over time, so did my partner Nina, followed by Jemima, followed by my mum (super star) and a variety of other people. We were able to establish that she was in fact prosocial to other dogs. Caution and vigilance would always be needed around small dogs, but Panda was finally found her forever home, where she has been now for five years or more. Panda was a challenge, and an extremely unusual one at that. She might well have been a challenge too far for another rescue and ended up, perhaps, as a 'sponsored' dog, kept safe and well in a shelter, but beyond the hope of rehoming. There would have been no shame in that. But Jemima is a believer that every dog deserves a chance and Panda has repaid that confidence. And, for the record, she and the Postie who delivers to her forever home are as if they grew up together. Win!

CHAPTER 14

SIGNIFICANT OTHERS

A PROBLEM SHARED IS A PROBLEM SOLVED

When I worked in the entertainment industry as a sound engineer, it was taken as read that a production was going to need a village to work on it if it was going to look good, sound good and get to completion on time to the satisfaction of the client and the all-important audience. Depending on the scale of the thing it could involve large numbers of people. It would need production people to come up with concepts and liaise with the client. They would be likely to employ logistics people, who would organise when, where and how people were going to get to the event. Caterers (may their lives be forever happy!) might need to feed tens, if not hundreds, of hungry people. (A good catering crew is worth its weight in gold.) Production managers would coordinate the different technical departments both before, during and after the event.

If you are the head of the department, as I sometimes was for the sound department, you would need to attend one or more production meetings. You would need to work alongside the lighting department, the VT department, the staging department and the rigging department, all of whom had their own priorities, but all of whom understood the need to collaborate to a common end. It was crucial that everyone got on together and had that common objective in mind, as well as recognising one another's expertise in their chosen profession. That meant not just making way for one another, but actively helping each other too. Things just wouldn't get done otherwise.

I came to dog training with that same ethos front and centre. It hadn't occurred to me that it might be quite such a solo enterprise. Often it is, of course, but I was and still am open to it being a village affair. You need to be able to identify behavioural indicators in dogs that may steer you towards a cause that you don't have knowledge of or expertise in. A production manager on a gig isn't necessarily going to know what kind of speaker and how many the job is going to need, but they will know someone who does. That's how dog behaviour and training should be. If you need a veterinary referral, you should be able to get one and with someone who is going to talk to you as an equal. They might notice something that means a neurologist needs to chip in, or the vet may consult a colleague in their practice who has a particular interest in nutrition.

My point is that as a professional practitioner you should be prepared to say 'I don't know' if the question is outside your field. You should also be able to find someone who can reliably answer the question for you.

Yet without a fair degree of research it can be hard to know exactly who can help you do what and it can be tough knowing who to trust for what. The field of the dog trainer, particularly, is an unregulated and often uncharted territory, particularly for a dog guardian who may already be feeling worried and stressed. I have heard all too often of dog owners being blamed and shamed by trainers for their dog's behaviour, rather than helped.

There are even more disturbing stories of clients with fearful dogs who have sought help and then been told by the trainer they approached that they would have their dog removed as 'dangerous' unless the owner agreed to a training programme at an exorbitantly expensive cost. This is not professional help, it's extortion.

The various trainer trade bodies and accreditations don't necessarily help owners navigate this minefield, presenting an impenetrable alphabet soup of options to the unwary and uninitiated. In recent years, the drive within the industry appears to have been towards ever greater fragmentation, rather than uniting in the cause of good professional standards and making the right help available to dog lovers.

Vets are, of course, properly regulated and rightly so. They'll be highly qualified in the health of your dog, but they may have little formal education or training in dog behaviour. It is the veterinary behaviourists and clinical animal behaviourists who have specific behavioural training. Dog behaviour actually features very, very little in a veterinary degree, often only given a few days. This is not altogether surprising given the scope of what they have to cover – from small pets to farm animals. Nevertheless, the gap in knowledge has been recognised as an issue. In fact, as recently as 2023, the Dogs Trust in the UK launched the Dog Friendly Clinic Scheme with the British Veterinary Behaviour Association. This is doing valuable service in offering information on dog behaviour, how dogs learn and addressing common misconceptions around dominance and guilt. The scheme was established with the aim of making 'the provision of veterinary care the most comfortable experience possible for everyone involved'. We can all get behind that.

So far, so complex. But when the professions collaborate well, the results can be life-changing for the dog and the people. In this respect, there is no relationship outside that of a trainer with their client that is more important than that between a trainer and a vet or vet behaviourist. I'm sharing here a story that highlights that point. Dora's case needed contacts, friends and colleagues and an ability to gather appropriate information, but the results have made it all worthwhile. After Dora's story, Sara, the vet behaviourist involved in identifying the cause of Dora's problem, talks a little about how you can set yourself up for success in your choice of professional help.

SIGNIFICANT OTHERS

DORA'S STORY

Dora is Tim and Ella's first dog and they were invested in her from day one – literally. Friends of friends had a working cocker spaniel who was having a litter and so they were kept in touch from the day she first appeared. I've seen the photos and, trust me, she was (and still is) adorable. They counted down the days until they could bring her home and took two weeks' puppy-ternity from work so they could settle her in. They played with her and taught her silly tricks. She was perfect.

Then lockdown happened. Travel and social contact were severely restricted, but Team Dora put the time to good use. Working from home meant it couldn't completely be a puppy exclusive timetable, but that was fine and she adapted to having a snooze in her crate while they worked.

As first-time puppy parents, there was a learning curve around socialisation, toilet training and all the things that get you off to the best start. But it all looked promising and when lockdown rules relaxed, it became evident that Dora was a social butterfly. Everyone, be they person or dog, was a potential new friend.

Then things started to change. At about six months old, some issues began to emerge. Dora was often lethargic and grumpy – a million miles from what you would expect from a young spaniel. The vet investigated and eventually found some digestive issues that were causing her pain and, unsurprisingly, making her deeply unhappy.

Once this was resolved, Dora went back to being more like the puppy they had welcomed into their home some months before. When she came into season, they'd see some post-season grumpiness, but, they guessed, that was just the way things were with Dora. It was manageable. There were a few minor resource guarding issues, and she was inclined to put teeth on flesh, but Tim and Ella got stuck into researching the problem and soon had it under control. Pats on the back all round. None of this was out of the ordinary for a developing young dog.

It wasn't until a month or so after Dora was spayed that things started to get worse again, and this time, there was no obvious medical issue. Training collapsed. Dora's resource guarding went through the roof and even normal activities could provoke aggressive behaviour. Her acquired bite inhibition deteriorated. Biting had become more frequent, the bites harder and it was becoming, frankly, really quite scary. The triggers for it were completely obscure. Even just standing up from a chair could cause growling and lunging.

She was super-sensitive to noises. Just the crinkling of the washing-up sponge could lead to a snap. Ella and Tim were very bewildered and feeling increasingly guilty. They knew one of Dora's littermates and she was reported to be doing well. Could all of this be their fault?

* * *

Before we go on with this story, I want to pause to say a word or two about guilt. In my experience, guilt is very common among dog parents who are faced with the terrifying realisation that they have a dog who reacts aggressively to certain people, dogs, things or – more problematically – to apparently anything at all. I often hear self-doubt and anguish from people who feel that, somehow, they must be responsible. But if you are in this place, you don't have to feel guilty. It could happen to anyone and the very fact that you are worried and seeking help to resolve it means you should give yourself a break. You have my admiration and recognition for doing a good thing. Kudos to you. There's no room for the blame game.

Let's get back to Dora. Things were becoming increasingly hard to manage. Tim describes this period as Dora's Jekyll and Hyde phase, but they didn't realise that things were about to change dramatically for the worse. They needed help so they began picking their way through the minefield that is finding a professional dog trainer and, eventually, ended with me. Following an online consultation, I put together a plan and Team Dora got to it with gusto.

But there was still something niggling in the back of my mind. Resource guarding is broadly viewed by the profession as one of the easier behaviour problems to resolve and, armed with a plan, I was absolutely convinced Dora's people could rock the training. Nevertheless, I was uneasy. Why had they seen this deterioration in her behaviour now? The three main reasons for training not being effective are: the wrong plan, poor compliance and poor execution. Well, the plan was a standard operating procedure for resource guarding, which has a significant track record, so not very likely. Yet they'd already succeeded in resolving the resource guarding and she'd regressed. Compliance? Given the engagement I was getting from the team I was pretty convinced of their dedication to any training programme. They wouldn't have done half a job, but they had seen the problem come back. Execution, or the skill in delivery of the plan? Well, that absolutely stood up to scrutiny. They were really good trainers. Something just didn't feel right. I needed to clear away the cobwebs of the different elements of the problem and go back for another deep dive into Dora's history.

There was a hypothesis that was formulating in the 'little grey cells'.

We all jumped on another Zoom call and a few things emerged, including some curious behaviour with visitors and neighbours. For example, one neighbour who helped occasionally with walks would ordinarily settle Dora upstairs when they got back and Dora would dutifully go to her bed. But on one occasion, Dora had remained downstairs apparently reluctant to follow the routine. When the neighbour started back down the stairs, Dora wouldn't let him out and it took some imaginative distraction before he was liberated. Another clue was that Dora, who likes her toys, had started to shuffle them around and guard some of them. She was choosing enclosed spaces to settle in. She was taking and caching not only her toys, but all sorts of weird and wonderful things that she managed to get her mouth on, from watches to the top of a reusable water bottle. While digging isn't unusual in any dog – and spaniels can be masters – Dora was digging and circling excessively on furniture at times. If she hadn't been spayed a couple of months previously you might have thought she was going through a phantom pregnancy.

It was all beginning to stack up. I recalled a webinar a friend and colleague had given to students and graduates of The Academy for Dog Trainers, where I'd done my training. It was some months prior to meeting with Team Dora and was presented by Devon-based veterinary behaviourist Sara Davies. The subject was Persistent Phantom/Pseudo Pregnancy (PPP) in spayed bitches and its impact on behaviour in subject dogs. If the dog has been spayed within a particular time frame or point of her cycle, then post-spay bitches could still exhibit the symptoms of PPP and those symptoms could be a combination of physiological and behavioural indicators, or the physiological indicators could be absent altogether. In a 2023 paper,[1] Sara Davies outlined that signs of pseudo pregnancy could show within days of spaying with some typical physical or behavioural signs.[2] But even months after spaying, affected dogs could show aggression towards humans or other dogs, anxiety and sound sensitivity without necessarily always showing the typical physical signs.

I proffered the possibility up to Team Dora, who were bewildered by the suggestion at first. This was understandable given she wasn't showing any of the physical symptoms of phantom pregnancy, the most common by far being mammary gland enlargement and/or lactation.

But we talked through the behavioural signs and I sent them Sara's articles. It was worth a shot – what was there to lose? They booked an appointment with their vet, armed with two of Sara's articles and suggested it might be something worth considering.

Dora's vet was a little sceptical initially, which isn't too unusual, but Tim and Ella had so much evidence on her behaviour that he saw no harm in trying what was, in reality, a low risk fourteen-day course of medication that had the potential to turn three lives around.

I knew that improvements can be seen in as little as a few days (although it's still crucial to take the whole course) and I was practically pacing, waiting for the first update from Tim and Ella. I was heavily invested in this little dog now, too. The first report back was really, really positive. It was such good news. But I still wanted to track progress to see if there still might need to be some ongoing behaviour modification. Dogs can acquire new learned patterns of behaviour during the persistent phantom pregnancy period that continue even after treatment, though they can be more easily addressed.

So what has changed for Dora? Well, her play drive has increased and so has her prey drive. Chase games are the best and she loves playing ball, but Tim and Ella have to be vigilant about her taking off after urban foxes if she spots one. In many ways, all the behaviours that you might expect in a spaniel, such as flushing, chasing, retrieving that were absent prior to the treatment are now very much coming to the front. She's also more selective when it comes to playmates in the park, but then that isn't necessarily unusual for any dog as they get older. Now, if she takes random items, she does it to start a game or to encourage them to ask her to 'drop it' for a treat. She's so enthusiastic about this, that the ferocious guarder will now throw objects in their direction if asked to give them up.

While the improvements seen as a result of the treatment have been remarkable, we shouldn't make this into a fairytale. There are still a few things to be managed (which for the most part, is the simplest solution to lower grade problems) or to be addressed with behaviour modification. The issues, however, are mild by comparison. In an interesting development, Dora did bite Ella once post medication, but it was clearly accounted for as a redirected bite when Dora and a Dachshund got into a spat in the park over a ball. The bite was little more than a scratch and although it's only a sample of one, the difference in Dora's bite inhibition is striking here. While she was suffering from PPP she was inflicting puncture wounds.

Ella and Tim still take sensible precautions in certain situations, such as vet visits, because she doesn't tolerate anything uncomfortable. She can still be grumpy when tired, so they know to make sure to give her a little down time. But Tim and Ella have learned to read Dora well and can identify potentially 'hot' situations in good time to avoid them or train to resolve them.

The important thing is that their Dora is back. Relations between Tim and Ella and their once cuddly puppy had become strained almost to breaking. Now love and fun are once more the order of the day.

THE VET BEHAVIOURIST'S STORY

Sara Davies is a very well-credentialled veterinary behaviourist working in Devon in the UK. Sara is registered with The Animal Behaviour and Training Council (ABTC) as a veterinary behaviourist and clinical animal behaviourist as well as being a full member of the Association of Pet Behaviour Counsellors (APBC). She is a regular contributor of webinars to both the APBC and to Jean Donaldson's Academy for Dog Trainers, of which she is an honours graduate. She is also a Fear Free Certified Professional.

These days Sara's main field of focus is on preventative behaviour and training interventions, medical causes of behavioural difficulties and on the effect of sex hormones and neutering on behaviour, a too little explored area of study. She has a particular interest in persistent pseudopregnancy, which makes me lucky to have had her insights on Dora's case.

'When it comes to anything that is about sudden change in behaviour, I strongly encourage getting a medical rule-out. There may be a need for both a medical intervention and behaviour modification too. In fact, that's not unusual at all.

'There's a tendency to see veterinary involvement purely in terms of musculoskeletal pain. In other words, the usual suspects like bones, muscles and the various tissues that connect them. But often the reality is more nuanced. I see more gastrointestinal issues than musculoskeletal, while other sources of pain or discomfort, such as dental or ocular pain, are frequently overlooked contributing factors.

'People don't tend to notice dental pain until their animals go off their food. But in the run-up to that, the pain is going to have been quite severe. As an example, my little dog had some dental issues last year. She had the upper fourth pre-molar teeth, the carnassial teeth, taken out from either side. She needed them extracted because she got slab fractures in the enamel and she did go off her food, which is very unusual for her. I then noticed after the dental was done, she was like a spring chicken. So, even though it takes a lot for them to show us they are in pain, a departure from the "normal" can be hugely telling.

'Looking back, her behaviour had been changing over time as she became more distracted by the pain. It can be difficult, even for a vet, to see gradual change in behaviour. I put it down to ageing, but getting the dental work done took two years off her in terms of her energy and enthusiasm on her walks.

'Pain is used as something of a panacea for a lot of behaviour change but it probably only represents about 30 per cent of cases that vets see. Of course, vets don't necessarily get to see all cases of sudden behaviour change in dogs because it relies on the guardian noticing and thinking that a vet consultation might be worthwhile. I'm not ruling out pain when it comes to behaviour change by any means. Quite the opposite. It's an important message to get across that, if you see sudden behaviour change in your dog, or any behaviour change at all in an elderly dog, or a problem is not responding to behaviour modification training as it should, get a vet check. But don't wait to start some behaviour modification with a qualified trainer or behaviourist. The truth is that if you contact a reputable trainer or behaviourist about aggressive behaviour, they are going to ask you to get a vet check anyway.

'There is something important to say here about choosing a vet. Just like trainers, you do not have to stay with the first vet you find. Find a practice you like, then find somebody you like within the practice and cling to them like a limpet. Unless it's an emergency, you don't need to see a different vet because if they are not in on Wednesday next week, you can just ask, "can I come on Thursday then?"

'Read the bios on their website to see whether there is a vet or veterinary nurse that has an interest in behavioural issues and contact them first. Perhaps ask to speak to that veterinary nurse first, because they will know which of the vets is the best bet to send you to for that initial consultation.

'In the normal course of their degree, vets do relatively little on behaviour. I graduated in 1993 had just one day in 1991 with psychologist and animal behaviourist Dr Roger Mugford. Now, at most they get about a week. The newer vet schools are far more forward thinking: welfare is interwoven from day one and behaviour is a big part of that. But the problem is, the curriculum is so stuffed. I had to learn all about horses, cows, pigs, knowing I was never going to work with them. At vet school you end up being a generalist. You graduate as a GP mixed animal vet.

'Now, of course those of us in the behaviour world recognise that anxiety can play a huge part in how dogs behave. We know how important a relationship between professional disciplines is and how important it is to be clear that medication is only one (very important) part of the process – behaviour modification is needed too.'

CHAPTER 15

LIFE CHANGES

Dogs have their lives structured largely around our routines, relationships and the places where we hang out. They can't just decide for themselves that they want to raid the fridge because they're feeling peckish, walk into town to try out that new café or head off to the pub because there's absolutely nothing worth watching on TV. We make the decisions around where they live and walk, when they eat and who they get to meet. And dogs learn those familiar routines, people and places. They are part of how they understand the world and, on the whole, it helps make them feel safe.

But while the anchor points of familiar routines are important to dogs, that doesn't mean they're not up for new discoveries: anyone who's ever taken their dog on a holiday or scouted out a new local walk will know they can get pretty excited about it. Like humans, it isn't that they don't like change per se. It's more about them understanding what it means for them and the benefits it might bring (or not).

When we moved home from England to Scotland, Ripley was pretty anxious about the whole process – and who could blame her? We did the move ourselves over a couple of weeks. Favourite pieces of furniture disappeared into vans, as did I. This was worrying. There was a general air of busy-ness and stress. Even when the move was complete, for the first few days she was uncertain and a little clingy. There were some familiar smells from familiar furniture, but to a super sensor nose, there were also lots of strange ones, including, very likely, those of the previous resident dogs, despite the thorough cleaning the last owners had done. But when she realised that ball on the beach was going to be a feature of her everyday new life, that put things in a whole different light. This was, without question, a change she could get behind.

On the whole, dogs can be pretty adaptive, which is just as well as in their time with us it's likely that they will experience some significant life changes: new homes, new partners, births and deaths, break-ups and departures to university or new jobs, new pet companions, even, perhaps, new people to take care of them. And things will change for them, naturally, as they get older and so do we.

While the blanket 'dogs don't like change' adage isn't universally true, they do like a routine that they can rely on and there will be times when we have to shake that up. Clearly, we can't try and get them on board with a PowerPoint change management presentation complete with appropriate bullet point slides on 'Your part in our success' and 'What this means for you', followed by questions. But we can help them access their powers of adaptability.

THE SHOCK OF THE NEW

Certain elements of predictability are more important than others. They want to know where their next meal is coming from, when there's going to be a walk or a play. But they don't need routines that are too obsessively rigid. They should be able to accommodate some flexibility for when the unexpected happens, for example, you're stuck in traffic on the way home and mealtime is delayed. Some dogs might try and persuade you it's the absolute end of the world if you don't get up and out for a walk at precisely the same time every day or they don't get their dinner exactly on schedule, but it isn't. A little light variety and give and take helps them build up their ability to adapt.

But when it comes to big changes in their lives, such as new partners or new babies, they'll be at their adaptive best if the unfamiliar is eased into their lives and, in the finest tradition of Pavlovian learning, the new reliably predicts something good for them.

If a new baby is on the way, for example, it's a good idea to establish the new routines and rules in advance of the new arrival actually arriving (if they're going to be excluded from rooms they've previously had access to, for example, or encouraged to sleep in a crate). If they can get used to baby furniture, gates, walking alongside a pushchair, unfamiliar smells like baby lotion, and even the weird sounds of baby cries or gurgles by playing recordings at low levels, the path will be smoothed. Even better, if they associate all this strange new world stuff with positives for them, like getting a treat or a Kong toy when they're on the other side of the gate, the shock of the new arrival will be less bewildering and destabilising. At least some of the unfamiliar will have become absorbed into their everyday life.

If the newcomer isn't a baby but a new dog, they might be welcomed into the household by the established resident with open paws, but there is absolutely no way in advance to know. It is worth considering the balance of power and personalities in any new relationship. A confident, mature dog may well have no trouble laying down the rules of the household to a naive, younger newcomer, whereas an old-timer that's been accustomed to an easy, gentle life as a solo dog may find a large, boisterous adult incomer way too much to handle. In these cases, the 'new lease of life' theory needs to be treated with some caution and commonsense.

Whatever the balance, the existing dog will need, ideally, to have had the chance to get acquainted with the newcomer on neutral territory before being invited straight on to their home turf to share their space and stuff. A period of adjustment is absolutely to be expected – the existing dog will, perfectly reasonably, be concerned about what this means for them and the safety of the things that matter to them: food, toys, sleeping places and you. So it's back to first principles of learning again: the new arrival must not mean loss for the incumbent but predict good stuff. There has to be, at the very least, toys, games, food, treats and attention enough for everyone: what the Ohio State University College of Veterinary Medicine describes as 'creating an environment of plenty'. It's worth saying, however, that even that doesn't accommodate for the fact that some toys may be more favourite than others, or that some dogs may want to guard the whole resource, any loss feeling like a threat to them, irrespective of how abundant the supply may be. Occasional spats are normal, as they are in any family, but if the newcomer predicts, at worst no loss and at best a gain to the dog in residence, then they should rub along nicely.

When Ripley arrived as a puppy, our Rottweiler, Murphy, established himself early and easily as the fun uncle. Given the disparity in sizes, we were extremely cautious to begin with and made sure to keep them separate when unsupervised and to run through the training process of making sure that the presence of the tiny fat-bellied pup predicted good stuff for Murphy. (I still have the video on my phone of Murphy playing the game of 'see the puppy, get a piece of chicken', while Ripley waddled around him, cautious, but curious. It still makes me smile.) Taking the time to make sure Ripley meant benefit, not loss, was a sensible precaution. But, in truth, Murphy never seemed to see her as a threat to his resources. No, as far as he was concerned, she was his apprentice-in-mischief and welcome to the team. He'd break off small corners of his treats and slide them over to her. And when a visiting dog snapped at her and she ran, terrified, to hide under a table, he knew immediately what to do. With a firm whack of the paw, he opened the door to the downstairs cloakroom, eased the loo roll off the holder, took it to where she was hiding and rolled it to her. Because, as every fond uncle knows, nothing makes an upset cockapoo puppy feel better than the opportunity to shred a loo roll all over the floor.

Our Cocker Spaniel, Woody, took longer to come round to the new housemate, which saddened us a little as he'd always been sweet-natured – the original sociable dog, loving the world and everyone in it. But it was understandable. He'd been suffering from an unpleasant skin condition that was only diagnosed too much later as the first symptom of lymphoma. Feeling uncomfortable, he really didn't want to engage with this bouncy, bumptious youngster. But even he succumbed to her charms eventually. This was, perhaps, because of the generous supply of training chicken available whenever she was in proximity. But, I suspect, it may have been more to do with the fact that, when she was finally exhausted with puppy play, she would tentatively climb on to the sofa next to him and go to sleep, and the thick, soft, curly fur of her rump made an exquisitely comfortable pillow. Like I said, change is OK if it comes with benefits.

However much it seems likely that your existing, sociable dog will welcome a new chum into the family, the time spent setting the relationship up for success is always well spent. Those same rules apply to when you're introducing a new human partner. However great you think they are, it makes sense to make sure your dog sees them as the predictor of all good things: treats, games, fun walks and outings and not just a drain on the attention resources of their favourite human. If a new romantic attachment can't do that, what on earth are you thinking of?

GREEN-EYED MONSTERS

Which brings us neatly to the knotty problem of whether dogs, faced with a rival for the affection, attention and bounty of their human, feel jealousy. You might have the sort of dog who will always try and get between you and any other dog that wants to come up and say 'hello'. Or the kind that wants to muscle in between you and your partner if you're curled up on the sofa. That seems like jealousy, but is it? Well, you are not the first to have wondered if dogs get jealous and some have tried to find out.

A study carried out by the University of California in San Diego looked at the reactions of thirty-six dogs when their owners ignored them and interacted with different things.[1] If they read aloud from a children's pop-up book or lavished attention on a bucket with a face painted on the side, the dogs were pretty indifferent. (Whether they quietly worried about the sanity of their human, apparently trying to get up close and personal with a painted bucket, we do not know.) But when their owners petted a realistic-looking stuffed dog toy, complete with bark and wagging tail, it was a whole different story. The dogs touched the stuffed toy, or tried to get between it and their owner, twice as often as they did with the bucket, and far more than with the book. A quarter of them snapped at the stuffed rival, while only one had done the same with a book or bucket.

Another study by the University of Auckland found that dogs reacted to their human engaging with a stuffed dog and tried to intervene, whether they could actually see what was going on or not.[2] If human and stuffed rival disappeared behind a barrier to commit their act of dastardly disloyalty, the dogs reacted in the same way. This led the researchers to conclude that 'dogs can mentally represent jealousy-inducing social interactions'. Oh, my.

That said, is this jealousy as we humans would know it? Certainly, there's no evidence that our dogs brood and simmer over the attention lavished on a rival. It seems to be very much in the moment, or 'primordial' as the researchers characterise it. And while they express their displeasure at the 'rivals', it would appear they don't hold it against us – they won't be cutting up our suits or keying our cars if we happen to give a pet to the poodle in the park.

Since we pretty much control access to the resources in this relationship, it is hardly surprising that our dogs feel the need to protect a valued social relationship from outsiders. We are the gateway to the good stuff and they want to protect that access. Is this jealousy? Or survival? I'm not sure. If you pushed me, I'd say I'm not sure the evidence presented is compelling enough on this one and I'm inclined to direct the jury to acquit our dogs of the charge of any green-eyed monstering.

RUBY AND SADIE'S STORY

Carla and her wife Beth had done something that, to my mind, is hard to beat on the compassion scale. They had adopted, and thereby rescued, an elderly dog. A small, almost toothless little sweetheart by the name of Sadie. Carla and Beth already had a dog. A Hungarian Vizsla called Ruby. Ruby was only young (she still is) and she loves life. She loves her people, her walks, her play and her food. Lamp posts? Excellent. Beaches? Top notch. 'Let's get to it people, things to do!' There will be Vizsla people nodding in recognition, no doubt. And Ruby with other dogs? Generally, fine by all accounts. There had been the odd non-injurious scuffle, but nothing that would be considered a problem: all teeth and posturing, but nothing beyond that. She usually liked to play with other dogs, which stood them in good stead for an adoption. There was no reason to suppose that things wouldn't go anything other than smoothly. Work and home routines made it all possible.

Sadie would get comfort and love in her twilight years and Ruby would get a bit of canine company.

Perfect!

Things went fine in the early days, but after a little time, while there wasn't a wild swing in the direction of trouble, Ruby had started to snap at Sadie. In the latest incident, Ruby had given Sadie a cut to her eyelid and a shallow puncture close to her eye. On a scale of one to five, this wouldn't rank as serious by any means. It's easy to catch an eyelid or a lip in the process of posturing during a spat and the tissue around the face is soft and vulnerable. In that sense, Ruby still hadn't acquired too bad a rap sheet. Nevertheless, it was a worry and, as far as Sadie was concerned, not something you need to happen when you are thirteen and expecting a nice retirement home by the sea. Carla and Beth decided to look for help, wisely seeing it as a 'let's nip this in the bud' moment. I gave them some things they could do to get things started over the phone while we sorted out a slot when I could visit them in person. They only lived a couple of miles away.

The common name for issues between dogs in a household is 'sibling rivalry'. It isn't the most accurate descriptor of what is happening. First of all, the dogs in question do not need to be blood relatives and it isn't motivated by something as abstract as status. But it serves as a pithy name for a more complex problem.

Generally, by the time things have got problematic between dogs in the household and sibling rivalry has set in, escalating to more frequent and injurious assaults, it's really quite difficult to identify the trigger. People usually talk about it as 'unprovoked' or 'out of the blue'. 'They just don't seem to like each other.' When sibling rivalry has become established, that assessment of 'they just don't like one another' has become true. They have sensitised to one another's presence. Where once a location or particular set of circumstances generated a problem, they're now simply predicting that a fight is likely to break out, which means a 'let's just get this over with now' attitude prevails. Oh, and to add some icing to the cake, it's statistically worse, both in terms of frequency and outcomes, with girls.

All that said, we were in a good place with Ruby and Sadie. Carla and Beth had been remarkably proactive. They weren't about to let this behaviour bed in and when I met them I was in no doubt that they were going to get it done. Failure wasn't an option. It was clear from the offset, too, that they had a great sense of timing and that is always important in any training protocol. Whether you are training for behaviours or emotions, your dog needs to make a connection between one thing and another. Beth and Carla had not only made a start on training but had been vigilant about management too. They had deployed exercise pens where necessary to keep the dogs apart and were acutely aware of potential hotspots or situations that might trigger Ruby's reaction. That's not always an easy thing to do, but this early intervention meant we could make the plan of action truly efficient.

The original, and in this case, current motivating features were still clear, thanks to their prompt action. It's almost always the case with sibling rivalry that it has some form of guarding in its history. It may be food, places or people but it frequently accounts for the progression into 'dislike' later on. (That is assuming that there isn't a fear of dogs issue, but it's rare that people attempt to introduce another dog to a household with a dog already sensitised to other dogs. That has usually flagged itself well before the idea of a second dog enters the equation.)

I was pretty sure I could identify the issue with Ruby. I didn't actually need to witness it – I wasn't about to knowingly subject Sadie to that stress. The description of it was enough and I could see how the layout of the house worked in that regard. The point about resolving any behaviour problem be it guarding, fear of people or fear of dogs is that when you ask the question: 'Is my dog upset or not?' and the answer is 'yes', you don't need to see the problem rehearsed in order to start work on resolving it. In an ideal world (and I know we don't live in one) you won't see the problem again.

The other big bonus with guarding behaviours is how well they resolve with standard operating procedures. If you had to have a behaviour problem that was about fear or anxiety, resource guarding is your guy. Follow the plan and you will generally get it done. That is where we were with Ruby and poor retiree, Sadie.

Sadie, not surprisingly, had started to get twitchy about Ruby being close to her. I'm not generally inclined to speculate, but I think it seems likely that because Sadie was finding things difficult to predict now, steering clear was the simplest option. It's important to stress that neither dog was at fault here. This wasn't an exercise in blame allocation, but conflict resolution. I wanted to establish a mutually beneficial situation for both parties and ultimately end up with a harmonious accord. This meant creating a positive emotional response from Ruby towards Sadie. But it also needed everyone to be aware that Sadie wasn't just a target but a 'victim'. She will have developed some negative emotions towards Ruby and that needed addressing, too. That was the least we could do for her.

There is a simple exercise we can use to this end. Ruby is tethered or held on leash by either Carla or Beth. The other person brings Sadie into the room also on a leash as far away as the room design allows. Sadie gets a treat and immediately afterwards, Ruby gets a whole party's-worth of treats. Sadie's attention is maintained with treats and Ruby's party continues until Sadie leaves. That was the cue to stop the party and make things pretty boring for Ruby. This set-up was repeated until Ruby showed some sign of pleasure at Sadie entering the room, at which point they could move a little closer with Sadie. The order of events is crucial here. Ruby must be aware of Sadie before the party starts. Sadie predicts treats; Sadie leaving means party stops. A (Sadie) predicts B (food). Oh, and just so we're clear, kibble or biscuits don't cut it in this situation. You are asking for something valuable so pay accordingly. Put your best trade forward. Whatever really floats their boat. Of course, if that's biscuits then fair enough. I'm not about to argue the point. We're all entitled to our own tastes!

MINE MINE MINE

I said earlier that the root of sibling rivalry, often long since clouded with time and mistrust, is frequently resource guarding. A valuable resource is defended from potential 'thieves'. The intention of the supposed thief doesn't come into it and nor does the availability of the resource. There may be more than enough to go around for everyone to enjoy some without depleting the supply for others. The guarder guards. That's what they do. The threat is to the whole supply. Period. You can look – from a distance – but you can't touch. Or look for very long for that matter. It's all MINE! Resource guarding need not be universal in nature either. The guarder may guard from people or dogs but not necessarily both. They may guard from some people but not others. Just because they guard toys doesn't mean they are going to guard your shoes, although they might. One bed may be more 'important' than another and so on. It doesn't have to make sense. Remember the 'why' doesn't matter. It's the 'what' is happening that matters. The important question is: 'Are they upset or not?' If 'why?' moves the dial, then fine, but it's more likely we may never know the answer, so it's just as well we don't need to know.

Ruby's concern was her people. She cherishes them, so any interloper, oldie or not, was a bit of a no-no as far as she was concerned. She didn't have to share before. They could be taken away at any moment. That would be a disaster not worth contemplating and it was making her very, very anxious. She didn't have a problem with other people near her people. In fact she is delightful with people if her reaction to me is anything to go by, but dogs, hmm, not so much.

There's a plan you can follow to deal with people guarding. There's always a plan, a sequence you can follow to get to the goal. It's always incremental and it's always about reaching one stage's goal before going on to the next. It has steps that take into consideration the specific challenges you face and the degree of difficulty they represent. The plan works because it follows the principles of how dogs learn.

A plan is what Carla and Beth followed and at my last check-in Ruby thought Sadie was fabulous. Excellent.

OUR EVER-CHANGING MOODS

The evidence seems clearer that our dogs can read and tune in to changes in our mood – not just in tone of voice, but body language and expression, too. They have, after all, evolved alongside us so they've had plenty of time to learn our 'tells'. The Universities of Lincoln and Sao Paolo ran a test in 2016 to see how well dogs could recognise human emotions.[3] Dogs were shown images and played the sounds of different emotions, both positive and negative. They spent longer looking at the images that matched the sounds, leading researchers to conclude that dogs use sensory information to get 'a coherent perception of emotion'.

It's not just sight and sound, either. They can actually smell when we're afraid or getting stressed out, according to a University of Naples study,[4] and this has an effect on their cortisol levels. The researchers also found that female dogs could smell happiness as well as stress, though the male dogs in their study seemed oblivious to 'eau de positivité'. I don't know what we should make of that.

It's important, though, not to feel guilty that our stress might be automatically transferred to our dogs. Provided we're not being constantly bad-tempered with them and we're engaging positively, playing and spending time with them, there's evidence that there can be positive modulation of stress between bonded humans and dogs with both finding benefits to their wellbeing. Together we can very much be a mutual support network.

LOSS AND GRIEF

Some changes are, of course, very final and it's clear that dogs feel grief and experience loss. We can't, of course, explain to our dogs whether an absence is temporary or permanent. If a family member, human or canine, has been unwell, a dog may well notice a change in smell or behaviour. They are also, as we've seen, adept at reading our emotions, which may give a clue that the absence of a valued human is not just a holiday, a work trip or a departure for university, but a loss for all time.

But the feeling of grief that dogs experience isn't just a mirror of our own personal emotions. The University of Milan looked at how dogs mourn the loss of canine companions.[5] They found that a huge majority of dogs played and ate less and slept more. This happened irrespective of the level of attachment between the human and the dog that had died, which indicates the surviving dogs weren't simply projecting human grief. (It also, to nail a quick myth here, made no difference if the dog had seen the body of their late companion.) So, yes, dogs grieve their own losses, not just the ones we feel intensely about ourselves.

The stories of faithful dogs, mourning the loss of their human guardian are the stuff of legend. One of the more celebrated is the story of Greyfriars Bobby. Bobby, so the story goes, was a Skye Terrier, the constant companion of an Edinburgh nightwatchman in the late nineteenth century. When his human died of tuberculosis, Bobby kept vigil beside his owner's grave in Greyfriars churchyard for fourteen years, despite initial efforts by graveyard staff to move him on. His loyalty became famous in Edinburgh and crowds would gather to see him. The Lord Provost bought him a collar and paid his dog licence – it's good to have friends in high places. When Bobby died at the grand old age of sixteen, he was buried in Greyfriars, close to the human that he loved so loyally. There's a statue of him just outside the churchyard.

Sadly, the truth of the story has been called into question recently by historian Dr Jan Bondeson of Cardiff University. It seems that the Greyfriars Bobby legend might, in fact, have been an early exercise in marketing and PR. Bobby, so this alternative story goes, was a stray, looked after by the churchyard curator, James Brown, who quickly cottoned on that telling visitors to the church the story of the dog lying faithfully on his master's grave, earned him tips. Taking a walk with Bobby every lunchtime to the restaurant of a local businessman so Bobby could have a meal, and the restaurateur could take advantage of the custom of the accompanying fans, proved even more lucrative. Indeed, it was so good for business that when the original 'Bobby' died, he was quietly replaced so the profitable venture didn't come to an end.

I'm not at all disheartened by the possibility that Greyfriars Bobby didn't spend most of his life in mourning. I would like to think that even the most loyal, loving dog would, with encouragement, have abandoned his vigil long before fourteen years had passed and adjusted to life with a new human, content to take the occasional walk to the graveyard to pay his respects to his former guardian. Dogs undoubtedly feel loss and grief, whether because of death or relationship break-up, but they can adapt and help us to adapt, too.

But let's lighten the mood here. Not all absences are permanent. And nothing, but nothing compares to the unbridled joy of a dog who has been feeling the loss of a favourite human only to be reunited when a student returns from university, a soldier from deployment or a dog trainer from a work trip away.

SLOWING DOWN, BUT NOT OUT

The other unavoidable change we face with our dogs is getting older and it's all the harder to see because they age, in many ways, just like us, but much faster. The different parts of our bodies that get creaky or don't function so well as we get older, whether it's our joints, our heart, our brain or our gut, are equally affected in our dogs. Like us, dogs slow down when they age – losing muscle mass and sometimes developing arthritis. They might get more irritable, too and less tolerant of boisterousness from other dogs – almost certainly some will be woofing disapprovingly under their breath about the behaviour of 'pups of today'.

Elderly dogs can also suffer from a form of Alzheimer's, canine cognitive dysfunction. Dogs affected can be confused and might not recognise their human housemates or other familiar people or respond to familiar cues and commands. They might get lost in the house or be unable to navigate their way out from under furniture or the corner of a room.

Their sleep cycles can be completely disrupted so they pace around all night and sleep during the day.

There are things that we can do to help our dogs 'age well'. Staying active in body and mind with exercise and puzzle games helps, as does staying active socially. One study by the Dog Aging Project, a long-term collaboration between the University of Washington and Texas A&M University, found that social companionship from adult human companions and/or other dogs tends to keep them healthier in their older years.[6] It seems that what is good for the ageing dog is not dissimilar to what's good for us as we get older.

In fact, there are so many parallels between dog and human ageing that the Dog Aging Project, which is funded by the US National Institute of Health, is carrying out a whole programme of different research studies. They have enrolled 50,000 dogs into their programme to track their health and life experiences as they age. The aim is to uncover the keys to a long, healthy lifespan for our dogs and make discoveries that will benefit both the dog population and potentially have implications for people, too.

LUNA'S STORY

'She is beautiful, but often seems sad.'

All the enquiries I get from people looking for help with their dogs touch me in some way. But this one, particularly, reached through the wireless waves of the worldwide web and tugged at my soul.

When Deb wrote to me asking for help, it was not because their dog, Luna, was biting them or growling at them. She wasn't guarding the furniture or holding their limbs to ransom over a toy. She hadn't assaulted their relatives at Christmas or pinned the plumber to the corner of the kitchen with a menacing stare. I work with many people desperate in their need for help. But that simple sentence pierced my heart: 'She is beautiful, but often seems sad.'

Deb and Ed wondered if she was grieving the loss of their older dog, Bonnie, who had died a year previously. From the day she arrived in the household as a puppy, Luna had hero-worshipped Bonnie. The compliment wasn't entirely returned. Bonnie, aged seven, was initially pretty unimpressed with this ridiculous 'mini-me' with her endless attention-seeking antics and lack of respect for a dog's personal space. Couldn't a girl just enjoy a snooze in her own bed without someone muscling in? Gradually, though, Bonnie took on the mantle of adoptive Big Sis. She could be sweetly protective of Luna when other dogs were around and would give her a gentle wash when she came back from the vet. They hung out together, competitors in love of mud for eighteen months. But then things started to change.

With hindsight, Deb and Ed wondered if Luna knew Bonnie was seriously ill before either they or the vet did. To begin with she was diagnosed with a hip problem, and they'd take her off hopefully for hydrotherapy sessions, but she didn't improve. By the time Bonnie's tumour was found, Luna was already showing signs of losing her bounce and joie de vivre. She seemed down. If Luna had identified that something was wrong, it wouldn't be surprising – dogs are extraordinarily skilled at picking up micro signs of changes in behaviour and smell.

When Bonnie died, Luna's mood took a further downward spiral. Over the following months, in many ways, Deb and Ed found themselves mourning the loss of not one, but two dogs. Luna was hiding away, spending her time under beds or out in the garden. She actively avoided the places, like the kitchen sofa, where she and Bonnie had snuggled up together. The enthusiastic welcomes that ought to be part of life with a Labrador had gone. It wasn't totally bad news. Some of the regular routines and habits persisted and she would go on her walks and play in the garden with them, but everyday life with her had changed and it felt like a poignant cry for help.

I had a visit to London to make and so took the opportunity to visit Luna and her people on the way back to Scotland. Casa Luna was on the outskirts of York, which I think of as my hometown, having grown up there from the age of five and gone to college there, too. It's an easy place to love. I hadn't been back for a while and spent an evening walking the streets, marvelling at how much had changed since I was last there – and pondering the sadness of Luna.

Deb and Ed had done some necessary health rule-outs with their vet before my visit, which was great. Any sudden or dramatic change from a dog's normal behaviour should raise a health flag and so a vet visit and check-up are important. A clean bill established, I arrived armed, as I always am, with as wide a range of tasty goodies as I could, given the limitations of train travel.

The early signs of the first meeting with Luna were positive. Most self-respecting Labradors who don't suspect you of malign intent have their food price and Luna was no exception.

Although a little cautious at first, her interest in my treat pouch climbed rapidly until it reached cruising altitude and speed. It was clear Luna was intent on doing whatever it took to persuade me to dip into the stash. So far, so good. She was prepared to engage and play the game.

That didn't mean I was going to dismiss the grieving hypothesis. Luna almost certainly missed her playmate, no doubt desperately. And I am sure that there was a lot of sadness throughout the household at the time of Bonnie's departure. I can't know if Luna understood the absence to be permanent or not. But since Bonnie died, one day had turned into two, a week into months and, before you know it, a year had gone by. And Luna was still sad.

Dogs are social animals who form special attachments to the members of their families. If that were not the case, departures wouldn't have them watching from the window for hours and those special welcomes at the door when we return at the end of a working day would not have the exuberance that they do. But we needed to get Luna to re-engage with her people and rediscover more of the joy in life.

Because I specialise in working with dogs with fear or aggression, changing emotional responses is a huge part of what I do. There's no hokum involved in this. The methods I follow are science-based and use tried and tested scientific principles. And behaviour science tells us that Classical, or Pavlovian, conditioning and the way it functions in relation to domestic dogs, does more than effect just physiological change. It can change emotional responses too. And what greater emotional change is there than to build and cement relationships between individuals?

Luna had, I suspected, learned a new routine since Bonnie's death. Grief had turned into behavioural normality and carved a deep groove into the pattern of her everyday. We needed to give her reasons to abandon her mourning and create new habits.

We started with some simple obedience 'games' that Deb and Ed could use to get her to spend time in the same room with them, rather than hiding away. It soon became apparent that Luna was up for earning her pay. 'Go to your bed' was her game of choice – so much so that she pretty quickly plundered my treat reserves, enthusiastically pitching herself into her bed almost from the other side of the room. Excellent job, Luna.

The other concern for Deb and Ed was Luna's tendency to stay out in the garden. Not surprisingly, they'd interpreted this as a reluctance to be in the house, another aspect of her hiding away behaviour. But it didn't strike me that way when I saw it.

Certainly, she wasn't coming back in. She would stand and stare, as if she was being asked to step off a precipice, rather than over a threshold. Yet she would find a toy in the garden to bring to you or chase a ball if you threw one for her. Remember the first question on the mental checklist for identifying the cause of a behaviour? 'Is the dog upset or not?' Well, in the garden clearly she was not. She was rolling around on the rain-drenched ground just like any mud-loving Labrador should. She would happily walk up to the door for a treat, but just wouldn't actually come in. Right. OK! If she's not upset, then something must be reinforcing the behaviour.

We could absolutely rule out any punishment history for coming in the house. There was no question of that. So what was motivating her behaviour? Remember that behaviour is economics: a profit and loss sheet. What gets me 'pay' and what does not get me 'pay'? A dog will do more of the former and less of the latter.

There were two things that Luna liked in this situation: people being outside and a tasty treat. If she held out long enough, she would get one or the other. Either someone would go out to her and she could play 'chase me games' or she got a tasty treat to lure her indoors. She was a smart cookie. 'Did someone say cookie? Don't mind if I do.'

The solution here was to shift the balance of profit and loss for Luna when it came to staying out in the garden. The approach was two-pronged. First, we needed to penalise failures to come in on the first request so that it now became loss-making, as far as Luna was concerned. That meant that if she failed to come in, the door would be shut. No play. No treat. Luna would come to the door and sit a few feet from the threshold with that 'show me the money' look that I've come to recognise in dogs who have pulled this swindle on their people before. Door closes. Rinse and repeat. After four or five attempts, she was coming straight in. Cue the reward party for Luna. Abba tunes, silly hats, confetti, a big cake ... Well, it was chicken, rather than cake, if truth be told, but the message in these situations is always: 'Don't be stingy at this point. Spend the big bucks.' More practice, more repetition would be needed, but we'd taken the first step.

The second prong of the strategy was to make outdoor play contingent on Luna coming into the house first. It sounds absurd, I know, but the idea was to teach Luna that to get what she wants, she must do the very behaviour that apparently rules out achieving her objective. Don't believe me? Here's how it works. I call Luna in from the garden into the house. When she does, eventually, turn up and come in, I briefly close the door, but then open it again and go out with her for a game of ball. Repeat. Over time, the duration between calling her and her turning up becomes shorter as she learns that this is, in fact, the fastest route to reinforcement. It takes regular practice and, in the short term, plenty of reward to protect all the training. But, in time, it becomes second nature.

LIFE CHANGES

Ed and Deb are also discovering more ways to engage her and get her back to spending time with them. Sometimes obvious training games are not the only way forward. Ed sits on the sofa and plays guitar to her, which she loves. Music is often overlooked as something that can bring dogs together with their humans – it's just one more thing we share.

Her progress is ongoing. There are times when she still seems to retreat into herself, but working in collaboration with her vet, we're hoping that she's on her way to happy.

CHAPTER 16

IN CONCLUSION

'If you talk to a man in a language he understands,
that goes to his head. If you talk to him in his language,
that goes to his heart.'

– Nelson Mandela

I suppose that it is stating the bloomin' obvious to readers of this book to say that dogs are remarkable animals. Personally, I could watch them for hours, be they sleeping or goofing around in the snow or freshly mown grass. They might be playing with one another or us. They might be dragging their hapless person back to a nearly missed sniff or a discarded takeaway. Even when they squabble there are things to learn from and about them.

They hold a fascination for me and I have a deep admiration for them. How have they come to be so integrated into our lives? A quite staggering number of us welcome them into our households and for most they are not simply 'the dog'. It's far more complex and wonderful than that. We could be forgiven for thinking as a novice dog guardian that it must be simple. We just need to scratch our way through those first, early weeks and we'll be right as rain. The rest is child's play, quite literally a walk in the park. Sometimes, perhaps even often, that is true. If it weren't there wouldn't be nearly half a billion pet dogs in the world. We would have long since thrown in the towel or run away laughing hysterically at the mere suggestion: 'No sir! You're having a laugh. Live with a dog? Are you completely mad?'

But we do it in our millions. We scour the internet for that canine Miss or Mister Right and even any warning labels attached don't deter us. The countless blog posts that describe how your routine will change and how your responsibilities will shift on to someone in your household whose needs will have to come first for a lot of the time are read cheerfully. If you are adopting a rescue dog those labels will, if the rescue is good, include the challenges that you might face. You look the objections in the eye and say, 'We can do this.' Often even before inviting them into your home, you've stopped talking about 'a' dog and have far surpassed 'the' dog. You are now talking about 'our' dog. Whatever it is that compels us to share our homes, our lives, our sausages with 'a' dog, it's got us and there's no letting go. You're in it for the long haul now and you are braced for whatever it may bring.

IN CONCLUSION

MAKING A LIFE WITH DOGS

Full disclosure: professionally, I rarely see completely happy dogs. I don't often see the dogs who are happy and contented enough, where the behavioural wrinkles are under control and probably don't matter anyway. But occasionally, I do. Sometimes I get calls about sociable dogs and their recall or watchdog barking or jumping up at guests and drowning them in slobber. The ones who pull like a train on lead? I like those. The solution is usually an 'aha' moment for people. And they are fun.

But if your business card says, 'specialising in fear and aggression', it's going to skew the sample pool somewhat. But let's be clear: there are millions upon millions of happy and contented dogs in the world. I don't know how close to the moon you would get if you stood them one on top of another, but a long way, I wager. I get to see a disproportionate amount of the others. The ones who are cowering from fear or biting out of the same motivation. I see the ones who are shouting: 'Get away from my stuff. I don't care if you gave it to me. It's MINE.' I see the ones who are fighting with the neighbour's dog or nipping the neighbour for that matter. Those dogs, the ones for whom and for whose people the lines of communication have broken down, are my drive and my purpose.

Ever since those sometimes terrifyingly confusing days with Thomson, when the blood dribbled off my hand and the challenge became shockingly real, I have been focused. As I graduated from Jean Donaldson's Academy it was 'difficult' dogs I wanted to help. I had graduated from helping 'this' dog to helping any dog who is trying, screaming out, to explain how they feel. And so for a lot of the time my field of view has been quite narrow.

But the purpose of this book wasn't to focus on only those people who have been calling in the wilderness. I wanted to speak to all of them, dogs and people, contented or wrestling with the odd challenge, in the hope of sharing something new, or rather, new to you. I wanted to offer something that will make the time you spend with your dog even more precious for knowing it. I'm not saying that my insight is completely 'new' or unique or innovative, because, in truth, it isn't; the science by which I live has been around for a long time, at least since the early twentieth century. But we live in a world that can generate a lot of white noise around facts, making it difficult to identify or even hear the truth in the first place.

The idea of 'pack leadership' or 'dominance' took hold because it allowed people to find easy explanations for things that were hard to account for. It was 'sticky' and 'pithy' and put us in charge, and, let's face it, control is something humans feel the need to have. Behaviour science has been waving from the wings for quite a while now, willing for people to pay attention. Meanwhile the illusionists wave their handkerchiefs and misdirect well-meaning, money-paying dog people to look the other way at a promise that cannot and will not be fulfilled. Evidence-based practitioners like me are getting better at putting our music on louder and making our commitment more obvious than the false promise, but we can still do better. We've had decades and it can only be our fault that we are still not the headline act for many people. This book is my part in changing that.

Learning is a common language. If we can understand how dogs learn and by doing so learn what really motivates their actions, that does a lot for them and for us. We discover what we have capacity for, how far our patience and endurance extends and just how much joy there is still to share with our dogs. We can find out what their capacity for that joy is and exactly what they are trying to communicate to us, using the only resource they have, their behaviour.

OUR SOCIAL CONTRACT

I advocate for dogs, for sure. In many ways I am as much a language teacher or an interpreter as I am a trainer. It matters to me that people are unhappy, frightened or desperate in the same way it matters to me that dogs are fearful and struggling. It is really only through the understanding of behaviour and what it is for, and the awareness of how learning works, that we can hope to live genuinely harmoniously with another species.

Being in partnership with dogs is what we are signed up to do. If it is not our original intention, as they gently lean into our affections it becomes a contract; a promise to another living being that, to the best of our ability, we will not let harm come to them.

Of course, the contract has a clause in it to bring happiness too. Those of you who have kids don't just ensure that they have their seat belts fastened in the car or that someone is waiting at the school gates for them when the end of day bell rings out. You want them to grow up knowing fun and adventure. There may be days, heading through holiday traffic to the beach with buckets and spades to explore rock pools when, if truth be known, you could do with just crashing out on the sofa after a particularly stressful week at work. But our responsibility extends beyond simply ensuring safety and, naturally, we get pleasure from that too. It's the pay-off for altruism. Is there a better feeling than seeing your child smile or your dog wag their tail?

Our ability to make meaningful connections with other animal species and their ability to return the favour is, perhaps, never more beautiful than the bond we have with dogs. Research suggests that they have behaviours that were evolutionarily developed just for us: from the rich variety of barks to the way they scan our faces for information and intention. It's a connection that has enriched our lives for millennia. If you've held your dog's stare, even for a moment, you can almost feel the history between us. That slight tilt of the head at a half-heard word that they've

IN CONCLUSION

heard often before. Familiarity for our dogs does not breed contempt. They have a skill for reading our movements for the tip-offs they offer for predictable outcomes. They learn our routines. For them familiarity is assurance, dependability. They are not always going to like what the telling stride or that glance towards the drawer where the car keys are kept means, but it gives them opportunity to respond.

But if they study our behaviour intently, it's not just for what that means for their own prospects. We are not just parts of their landscape, a reliable source of food or shelter. That probably is how this centuries-long relationship began. But what started as passing acquaintance and became a mutual utility has turned, ultimately, to love. When precisely we came to mean as much to them as they mean to us, we will never know. But mere familiarity can't account for that uncontrolled 'caution be hanged' welcome we get after an absence short or long. You know the one – the one the neighbour doesn't quite get, despite the fact that they, too, are familiar and known to carry a biscuit or two. The big display, the strike-up-the-band pageantry, that's reserved for family. And that is a deep connection, born of something intangibly more than mutual survival.

That is what I would like us, as human beings, to find and hang on to as hard as we can. That primal connection that we have almost told ourselves is wrong, weak, even destructive, not self-interested enough. I do not just want it for the warm and fuzzy feeling that it undoubtedly brings us; I want it for the ability it has to make everything else possible. Or at least, everything that makes living together happier and better for all of us.

Time spent with our dogs, whether they're chilled-out pups or more troubled souls, is always an investment worth making. What we have with our dogs is so much more than a passing, mutual convenience. Life with your dog may turn out to be something different than you first imagined, but if you enjoy that time together, simply be in the moment with them, whenever and wherever you can. It will be memorable and rich beyond measure.

Your dog, after all, is your friend, your confidant, your partner in crime. They're so much more than a dog. They're family.

ENDNOTES

CHAPTER 2

1. Kennel Club survey, 2023. Available at: https://www.thekennelclub.org.uk/media-centre/2023/june/canine-colleagues-are-in-demand/
2. LinkedIn survey for *People Management*, 2023. Available at: https://www.peoplemanagement.co.uk/article/1848131/people-management-poll-three-quarters-work-dog-friendly-office-hr-experts-say
3. Kurdek, L. A. (2009) 'Pet dogs as attachment figures for adult owners', *Journal of Family Psychology*, 23(4), 439–46, referenced in Adam Miklósi, *Dog Behaviour, Evolution and Cognition* (Oxford University Press, 2016). Available at: https://pubmed.ncbi.nlm.nih.gov/19685978/
4. Kluger, J. (2018) 'Why dogs and humans love each other more than anyone else', *Time*, 20 July.
5. Lord, K. A., et al. (2020) 'The history of farm foxes undermines the animal domestication syndrome', *Trends in Ecology & Evolution*, 35(2), 125–36. Available at: https://www.sciencedirect.com/science/article/pii/S0169534719303027

6. Miklósi, A. and Topál, J. (2013) 'What does it take to become "best friends"? Evolutionary changes in canine social competence', *Trends in Cognitive Sciences*, 17(6). Available at: https://pubmed.ncbi.nlm.nih.gov/23643552/

7. Payne, E. et al., (2015) 'Current perspectives on attachment and bonding in the dog–human dyad', *Psychology Research and Behavior Management*, 8. Available at: https://pmc.ncbi.nlm.nih.gov/articles/PMC4348122/

8. Albuquerque, N. and Resende, B. (2022) 'Dogs functionally respond to and use emotional information from human expressions', *Evolutionary Human Sciences*, 5(e2). Available at: https://www.researchgate.net/publication/367016074_Dogs_functionally_respond_to_and_use_emotional_information_from_human_expressions

9. Tuber, D. S., et al. (1996) 'Behavioral and glucocortisoid responses in adult domestic dogs (Canis familiaris) to companionship and social separation', *Journal of Comparative Psychology*, 110(1), 103–8. Available at: https://psycnet.apa.org/buy/1996-02664-011

CHAPTER 4

1. MacLean, E., et al. (2019) 'Highly heritable and functionally relevant breed differences in dog behaviour', *Proceedings of The Royal Society B*. Available at: https://royalsocietypublishing.org/doi/10.1098/rspb.2019.0716

2. Morrill, K., et al. (2022) 'Ancestry-inclusive dog genomics challenges popular breed stereotypes', *Science*, 376(6592). Available at: https://pubmed.ncbi.nlm.nih.gov/35482869/

ENDNOTES

CHAPTER 5

1. Fitzpatrick, Noel. *Listening to the Animals: Becoming the Supervet* (Orion Publishing, 2019).
2. Correia-Caeiro, C., et al. (2020) 'Perception of dynamic facial expressions of emotion between dogs and humans', *Animal Cognition*, 23, 465–76. Available at: https://pubmed.ncbi.nlm.nih.gov/32052285/
3. Rugaas, Turid. *On Talking Terms with Dogs: Calming Signals*, (Dogwise Publishing, 2005).
4. Hare, B. and Woods, V. 'What are dogs saying when they bark?', from *The Genius of Dogs* (Dutton, 2013). Available at: https://www.scientificamerican.com/article/what-are-dogs-saying-when-they-bark/

CHAPTER 6

1. Donaldson, Jean. *The Culture Clash*, (Dogwise Publishing, 2013).

CHAPTER 7

1. Howell, T. J., et al. (2015) 'Puppy parties and beyond: the role of early age socialization practices on adult dog behavior', *Veterinary Medicine (Auckland)*, 6, 143–53. Available at: https://www.ncbi.nlm.nih.gov/pmc/articles/PMC6067676/

CHAPTER 8

1. Behncke, Isabel. TED Radio Hour interview, 2015: 'What can Bonobos teach us about play?'.

CHAPTER 9

1. Dilks, D. D., et al. (2015) 'Awake fMRI reveals a specialized region in dog temporal cortex for face processing', *PeerJ*. Available at: https://peerj.com/articles/1115/
2. Pierce, J. (2018) 'Is your dog psychic?', *Psychology Today*. Available at: https://www.psychologytoday.com/gb/blog/all-dogs-go-heaven/201811/is-your-dog-psychic

CHAPTER 10

1. Harrow, Alix E. *The Ten Thousand Doors*, (Orbit, 2009).
2. Horn, L. et al. (2013) 'The Importance of the secure base effect for domestic dogs – evidence from a manipulative problem-solving task', *PLoS One*, 8(5). Available at: https://pubmed.ncbi.nlm.nih.gov/23734243/

CHAPTER 11

1. 2023 PDSA Animal Wellbeing Report. Available at: https://www.pdsa.org.uk/media/13976/pdsa-paw-report-2023.pdf

CHAPTER 12

1. Dittmann, M. T., et al. (2024) 'Low resting metabolic rate and increased hunger due to β-MSH and β-endorphin deletion in a canine model', *Science Advances*, 10(10). Available at: https://www.science.org/doi/10.1126/sciadv.adj3823
2. Beuchat, C. (2015) 'Health of purebred vs mixed breed dogs: the actual data', The Institute of Canine Biology. Available at: https://www.instituteofcaninebiology.org/blog/health-of-purebred-vs-mixed-breed-dogs-the-data

ENDNOTES

CHAPTER 13

1. Norman, C., et al. (2020) 'Importing rescue dogs into the UK: reasons, methods and welfare considerations', *Veterinary Record*, 186(8). Available at: https://www.ncbi.nlm.nih.gov/pmc/articles/PMC7057815/
2. St-Esprit, M. (2023) 'Are dogs and cats living longer?', VIN News. Available at: https://news.vin.com/default.aspx?pid=210&Id=11631825&f5=1

CHAPTER 14

1. Davies, S. (2023) 'Pseudopregnancy in spayed bitches – a preventable welfare problem', *Improve Veterinary Practice*. Available at: https://www.veterinary-practice.com/article/pseudopregnancy-in-spayed-bitches
2. Harvey, M. J., et al. (1999) 'A study of the aetiology of pseudopregnancy in the bitch and the effect of cabergoline therapy', *Veterinary Record*, 144(16). Available at: https://pubmed.ncbi.nlm.nih.gov/10343374/

CHAPTER 15

1. Harris, C. and Prouvost, C. (2014) 'Jealousy in dogs', *PLoS ONE*, 9(7). Available at: https://pubmed.ncbi.nlm.nih.gov/25054800/
2. Bastos, A. P. M., et al. (2021) 'Dogs mentally represent jealousy-inducing social interactions', *Psychological Science*, 32(5). Available at: https://journals.sagepub.com/doi/10.1177/0956797620979149

3. Albuquerque, N., et al. (2016) 'Dogs recognize dog and human emotions', *Biology Letters*, 12(1). Available at: https://pubmed.ncbi.nlm.nih.gov/26763220/
4. D'Aniello, B., et al. (2021) 'Sex differences in the behavioral responses of dogs exposed to human chemosignals of fear and happiness', *Animal Cognition*, 24, 299–309. Available at: https://link.springer.com/article/10.1007/s10071-021-01473-9
5. Uccheddu, S., et al. (2022) 'Domestic dogs (*Canis familiaris*) grieve over the loss of a conspecific', *Scientific Reports*, 192(1). Available at: https://pubmed.ncbi.nlm.nih.gov/35210440/
6. McCoy, B. M., et al. (2023) 'Social determinants of health and disease in companion dogs: A cohort study from the Dog Aging Project', *Evolution, Medicine, and Public Health*, 11(1). Available at: https://academic.oup.com/emph/article/11/1/187/7161464

ACKNOWLEDGEMENTS

There are some people to thank at this point. My partner, Nina, for her utterly invaluable help in making sense of my ramblings and her authorly contributions too. Without Nina I don't think I would have pulled a book from inside me into the light. Thank you, Nina. You are the best.

My Mum, Margaret Wooler, for being everything anyone could hope for a mother to be. For encouraging my love of furry things and everything she has done to allay my organisational inadequacies and step in to dog or cat sitting duties at the shortest of notices.

Ajda, our publisher, for firstly showing the curiosity to suggest that I might write this book and for her kind, friendly and cheerful reassurance throughout. What a star.

My thanks to all my clients for being hungry for knowledge and caring for their dogs so much that they went out in search of help in the first place. When things go awry it can be hard to appeal for help and I would encourage anyone to do it.

Particular thanks go to those people who gave permission for their stories to be told: Iola, whose commitment, love and dedication to her dog Winston and all of her dogs is an inspiration and the source of many happy memories.

Tim and Ella for spending so much time telling us about Dora's history and how they felt about the challenges they faced in such a transparent and honest way.

Deb, Ed and Luna. A heartbreaking story of loss and sadness. Sometimes what to do when you see such a change in your dog can be elusive. It can be almost paralysing – should we do nothing and wait or do something? Your willingness to talk to us about some painful things has touched our hearts. Thank you.

Jemima Harrison, who is (I've said it before and I'll say it again) a force of nature in the dog rescue world. Jem is a multitasker extraordinaire and her unwavering confidence in me is a humbling thing indeed. Panda was a difficult subject and my evaluation of the problem isn't common. It took a willingness to go with it on Jem's part and her own application to the cause. Brava.

Carla, Beth, Sadie and Ruby: When you contacted us and told us that you had rehomed a thirteen-year-old dog but introducing her to your young, happy Vizsla was not going to plan, I knew I had to help. That selfless, generous and kindly act of finding a place in your home and hearts for an oldie says everything about you.

Last but by no means least come all those people who contributed commentary and observation and those who

ACKNOWLEDGEMENTS

fact-checked me. I belong to a community of incredible people with extraordinary knowledge and experience who can be called on for collaboration or review. That's how it should be in our profession, as it is in so many others. We should be able to consult with one another and share our knowledge when asked for the benefit of dogs and their people. I feel very privileged to be a part of such a community and some have helped here. They include:

Jean Donaldson – my mentor and teacher without whom I would still be loading sound trucks at three in the morning. The Academy for Dog Trainers is one of the finest education programmes for dog trainers that has ever been devised. I owe a debt of gratitude to Jean for that and for her resolute friendship and guidance. Her confidence in my ability to write this book had a lot to do with its conception.

Sara Davies, BVMS, MRCVS, CTC, the vet behaviourist who generously shared her expertise in a field that I would otherwise not have been able to include. Staying 'in lane' is important if you claim professionalism and medicine is not mine. www.petsandtheirpeople.net

Jane Sigsworth MSc, CTC (hons) – as straight-talking a dog trainer as you are ever likely to come across, but one who has an immovable generosity of spirit towards people as well as dogs.

There are people who have read parts of these contemplations and made sure that I'm on a sound footing. It's easy to get the lines between unassailable fact and (albeit educated) assumption confused. Having some forensic or

MORE THAN JUST A DOG

even geekish brains on the case is invaluable. They are: Zazie Todd, Nick Honor, Denise Armstrong, Jane Sigsworth and Sara Davies.

INDEX

A

ABI (acquired bite inhibition) 157
The Academy for Dog Trainers 15, 247, 250, 286
activity shifts, in play 123
affection, importance of 115
ageing dogs 273–4
aggressive dogs 171–97
    stranger aggression 55
    tug as encouragement for aggression 96
Airedale Terriers 204
alpha rolls 89, 90–1
Alzheimer's 273
The Animal Behaviour and Training Council (ABTC) 250
animal welfare charities 217–18
anxiety 40, 253
    ears plastered back 73
    learning confidence from other dogs 93
    medication for 166–7
    resource guarding 188
    separation anxiety 93–4, 95, 154, 180
    Sophie from Romania 160–2, 163–9
approaching dogs 72–3, 74
arthritis 273

assistance dogs 126, 152
association, learning by 34–6, 188
Association of Pet Behaviour Counsellors (APBC) 250
attachment 25, 55, 277
    attachment figures 93, 148–9, 154–8, 169
    attachment theory 154–6, 162
    spending quality time with dogs 159, 169
attention-seeking 55
aversive techniques 179

B

babies 257–8
bad behaviour
    bad dogs vs bad owners myth 100–1
    rewarding bad behaviour 185–6
ball play 65–6, 96–8
barking 73, 82–3
    high-pitched barking 82
    low barks 82
    meaning of 25
    rapid barks 85
Basset Hounds 203
Battersea Cats and Dogs Home 218, 220–1
Beagles 203

behaviour
  changing behaviour 30–2, 38, 39
  dogs' ability to learn complex chains of behaviour 126–8
  learning by consequences 51
  nature and nurture 53–9
  understanding 29
  unprovoked 85
  what influences dogs' 44
  see also classical conditioning; operant conditioning, *and particular types of behaviour*
Behncke, Isabel 118, 119
Belgian Shepherd 204
Belyayev, Dmitry 21–2
Bernese Mountain Dogs 209
Beth 263–8, 269
bias, body language and 67
Bichons 205
the bigger picture 67–72
biting 79
  bite inhibition 92–3, 157, 244
  F.O.P. (fear of people) 181–2
  growling and 84, 85
  in play 122
Black Retriever X Rescue 228–32, 233–5
Blue Cross 18, 218, 219
Bobby 271–2
body language (dogs') 64, 65–76
  reading 180
  understanding the essentials 72–6
body language (human's) 269
body slamming 79
Bologneses 206
Bondeson, Dr Jan 272
bonds
  building 115–16, 152, 154, 155, 159, 169
  development of 148–9
  emotional support dogs 152
  play and 118, 119

Bonnie 275–81
Border Collies 55, 56
Border Terriers 204
boredom, and spite myth 95
Borzoi 202
Boston Terriers 205
bouncy gaits 80, 123
Boxers 205
brachycephalic dogs 207
breeds and breeding
  and behaviour 55–7
  breed-specific legislation 57
  early breeding 23–4
  purebreeds 201–8
  selective breeding 56, 143, 208
  traits 53, 55–7
British Veterinary Association 212
British Veterinary Behaviour Association 242
Broome, Reverend Arthur 217
Brown, James 272
Bulldogs 205

C

calm
  calming signals 75
  waiting for calm 109–10
Canine Compulsive Disorder 97
Carla 263–8, 269
Carville, James 31
Catherine the Great 206
cats 12–13
Cavalier King Charles Spaniels 207, 209, 211
Cellan-Jones, Rory 160–2, 163–9, 222
change, dogs' dislike of 255–7, 260
charities 217–32
Charles II, King 206
chasing 55, 122
Chihuahuas 205

INDEX

choosing the right dog 199–213
classical conditioning 35
    and emotional responses 37–8, 39, 278
    fearful dogs 44–9, 182
    predictability and 147
    thresholds 150
Cocker Spaniels 209
Cockerpoos 201
coercive relationships 32
collars, shock 128–9
Collies 203–4
comforting fearful dogs myth 98–9
communication
    between dogs 61–2
    breaking down communications 62–3
    face to face 64–5
    via body language 64, 65–76
    via play 77–81
    vocal 82–5
conditioning
    classical conditioning 35, 37–8, 39, 44–9, 147, 150, 182, 278
    counterconditioning 40, 99, 115, 151, 164, 178
    operant conditioning 32, 37–9, 49–51, 147–8
    Pavlovian conditioning 35, 37–8, 39, 45, 147, 182, 278
confidence, building 26, 104, 119
connections
    emotional connection 25
    importance of building 115–16
    meaningful 288–9
    mutual trade agreement 23–4, 32–3, 111
Conron, Wally 208–9
consent tests 71–2, 124
consequences, learning by 49–51
consistency, importance of 109, 133

contact, physical 26
    body handling 107
context 67–72, 73, 75, 84, 86
control 38
    who has control 32
cortisol 26, 270
counterconditioning 40, 99, 115, 151, 164, 178
Coyle, Diane 160–2, 163–9
coyotes 20
cross breeds, poodle 208–10
Cypriot rescue dogs 223

D

Dachshund 193, 196, 202, 249
Dalmatians 205
Darwin's Ark 55–7
Davies, Sara 242, 247, 248, 250–3
Dawkins, Richard, *The Selfish Gene* 53
daycare 18
Deb 275–81
desensitisation 40, 99, 105
    fearful dogs and 115, 145–6, 151, 178, 181, 182
    F.O.D. (fear of dogs) 181
    F.O.P. (fear of people) 182
    proximity sensitivity 178
'designer dogs' 208–10
destructive behaviour, reasons for 95
Dickens, Monica 208
diet, role in mental health 44
distress, signs of serious 150–1
DNA 54
Dobermanns 57, 202, 205
Dog Aging Project 274
dog cams 18
Dog Friendly Clinic Scheme 242
dog walkers 18
dogs
    domestication 19–23, 90
    early wolf-dogs 19–21, 23

fear of 180–1, 187, 266
genome 20–1
the modern dog 20, 56
dog's shelters 218
Dogs Trust 212, 218, 220, 242
domestication 19–23, 90, 143
dominance myth 89–90, 114, 286
Donaldson, Jean 36, 119, 250
The Culture Clash 14, 87
Dora 242–50

E

ears
ear positions 65, 81
pinned back against the head 69, 70, 73
economics, understanding the dog's 31–3
Ed 275–81
Einstein, Albert 77
elbow dysplasia 207
elderly dogs 273–4
emotions
dogs' recognition of human emotions 269–70
emotional connection 25
emotional responses 34, 37–8, 39, 44, 47, 278
emotional states 33
emotional support dogs 152
English Bull Terriers 204
epigenetics 57–9
escape strategies 150–1
exercise
dog walking 26–7
elderly dogs 274
experiences, introducing new 104
expressions, human 269
eye contact
emotional connection 25
'hard eyes' 74
and head tilt 76, 288–9
wolves and 21

F

face to face communication 64–5
facial expressions
dogs' 64–5, 79
human's 64
the play face 79
fearful dogs 35–6, 38, 40, 41, 42, 141–69
attachment figures 148–9, 154–8, 169
bases and bubbles 146–9, 152–69
comfort myth 98–9
and desensitisation 115, 145–6, 151
fear of strangers 116–18, 168, 181–2, 266
fear threshold 46, 149–56
F.O.D. (fear of dogs) 180–1, 187, 266
F.O.P. (fear of people) 116–18, 168, 181–2, 266
ignoring fear 98–9
learning confidence from other dogs 93
medication for 166–7
mistreatment myth 141–2
power of fear 45
recovery from 44, 46–7
rescue dogs 116–18
rewarding fear 185
safe spaces and bases and the trust conundrum 143–6, 148–9, 152–69
shutdown behaviour 150–1
sound sensitivity 143–4
spending quality time with fearful dogs 159, 169
and 'strong leadership' myth 91
triggers 46, 47–8
using classical conditioning to treat 45–9
feelings 33, 61

INDEX

fetch, as the wrong kind of play 96–8
fighting dogs
    dealing with 177–80
    in play 81, 122
    proximity sensitivity 176–80
fireworks 143
Fitzpatrick, Noel 61
Flat-coated Retrievers 206–7
flight, fight or freeze spectrum 42, 47, 77, 150–1, 173
Fluoxetine 167
F.O.D. (fear of dogs) 180–1, 187, 266
food
    and classical conditioning 45, 46–7, 48–9
    food guarding 90
    food rewards 110–12, 130–1, 132, 135–7
    training with treats as bribery myth 94
    and upset dogs 38
F.O.P. (fear of people) 181–2, 266
foster dogs 218
fox hounds 203
foxes 20, 21–2
French Bulldogs 56, 207
frustration, when on a lead 183–4

G

gaits, bouncy 80, 123
genetics 44–5, 47, 54, 58, 100
    epigenetics 57–9
    nature and nurture 53–9
genome 20–1
German Shepherds 56, 204
gestures 26, 133, 244
Golden Retrievers 208, 209
Goldendoodle 208
Greyfriars Bobby 271–2
Greyhounds 203
grief 270–3, 275–81
Griffon Bruxellois 207

grooming 107
growling 73, 83–5, 172
    in play 122
    punishing for 84–5
    signs of serious distress 151
guarding 266
    location guarding 90
    people guarding 269
    resource guarding 41, 42, 186–8, 244, 245–7, 259, 266, 268, 285
guilt 245

H

hackles 74
Harrison, Jemima 228–32, 233–5, 237
Harrow, Alix E. 149
Havanese 205
head tilt 76, 288–9
health checks 107
herders 203–4
hip dysplasia 207
hormones, emotional connection 26
hotels, dog 18
hounds 203
humans
    early relationships with dogs 19–23
    F.O.P. (fear of people) 116–18, 168, 181–2, 266
    human-dog myths 88–101
    reactive humans 189–92
    relationships with dogs 17–27
hunched posture 69, 70
hunkering 74
Huskies 205
hypersociability 21

I

identity, dogs as statement of 24
imprinting 45
inflationary spiral 130–2, 135–9
injuries, ball play and 97–8

305

innate behaviours 56
The Institute of Experimental Medicine 34
intentions
    intention signals 79–80, 84
    when good intentions go bad 125–39
intuition, body language and 67
Iola 193–7
isolation, social 26
Italian Greyhounds 205, 206

J
Jack Russells 204, 207–8, 209
jackals 20
jealousy 261–3
jumping up 73

K
Kennel Club 17
    Assured Breeding Scheme 213
King Charles Spaniels 205, 206, 207

L
Labradoodle 208
Labradors 56, 206–7, 276–81
Lakeland Terriers 204
learning
    ability to learn 54–9
    by consequences 49–51
    learned helplessness 150–1
Leipzig University 64
licking
    humans 73
    lips 67, 70, 75, 81
life changes 255–81
Lincoln University 64
lips
    licking lips 67, 70, 75, 81
    lip-curling 151
location guarding 90
loss 270–3, 275–81

Lucy's Law 211
Luna 274–81
lunging 38

M
Malteses 205, 209
Mandela, Nelson 283
MARS 123–4
Mech, Dave 89–90
medication, anxious dogs and 166–7
meta signals 78, 80, 123
Miniature Poodle 202
Miniature Schnauzers 205
mistreatment myth 141–2
mood, changes in 269–70
motivation 185
    dog's motivation 39
    for good behaviour 31
    for understanding behaviour 30–1
mounting 122
moving house 256
Mugford, Dr Roger 253
Murphy 131, 134–9, 205
    and Ripley 259–60
music 281
mutual trade agreement 23–4, 32–3, 111
muzzles 182
myths 87–101
    bad dogs vs bad owners 100–1
    breeds and the right dog 199
    comforting fearful dogs 98–9
    dogs as pack animals 92–4
    dominance myth 286
    pack leaders 10, 27, 88–91, 114, 286
    quick fixes 99–100
    spite myth 95
    training with treats as bribery myth 94
    tug and fetch as the wrong kinds of play 96–8

INDEX

N
National Canine Defence League 218
nature and nurture 53–9
Naturewatch Foundation 211
negative associations 128
new, shock of the 257–61, 263–4
new dogs, introducing 258–61
Newfoundlands 202
Nina 62
    and Murphy 135–7, 139
    and Panda 237
    and Thomson 15–16, 145–6
Nova Scotia Duck Tolling Retriever 202

O
obedience behaviour training 94, 108–10
Ohio State University College of Veterinary Medicine 259
older dogs 273–4
operant conditioning 32, 37–9, 49–51, 147–8
Oscar 12–13
oxytocin 26

P
packs
    dogs as pack animals 92–4
    pack leader myth 10, 27, 88–91, 114, 286
padding 151
pain, and behaviour change 251–2
Panda 232–7
partners, introducing new 261–3
'passive' training 159–62
pastoral breeds 203–4
patience 40–1
Pavlov, Ivan Petrovich 34, 38, 44
Pavlovian conditioning 35, 37–8, 39, 45, 147, 182, 278
paw raises 81, 123

PDSA Animal Wellbeing PAW Report 191, 210, 222
pedigree dogs 201–8
Pekingese 205
people
    F.O.P. (fear of people) 116–18, 168, 181–2, 266
    people guarding 269
People Management 17
Pepys, Samuel 206
perseverance 40–1
Persistent Phantom/Pseudo Pregnancy (PPP) 247–9, 250
Pet Advertising Advisory Group 212–13
Pets4Homes 209
phobias 35
physical contact 26
pinning down 79, 122
play 92
    absence of play 121
    bonding with 26, 118, 119, 167, 168
    communication via 77–81
    dog play 120, 121–4
    and fearful dogs 162
    importance of 96–7
    MARS 123
    meta signals 78, 80
    pauses in 81
    play bows 67, 79, 123
    the play face 79
    power of play 113–24, 162
    puppies and 92
    as a reward 112
    self-handicapping 81
    tug and fetch as the wrong kinds of play 96–8
pointers 202
Pompadour, Madame de 206
Poodles
    Miniature Poodle 202
    poodle cross breeds 208–10

posture, hunched 69, 70
practice 40–1
predatory behaviour, in play 122
predictability 147–8, 257
prediction, dogs' 126–9
pregnant dogs, impact of traumatic event on litter 142
prenups 18
prosocial dogs 73
provocation 172–3
proximity sensitivity 174–80
Pugs 205, 207
pulling dogs 109–10
puppies
    body handling 107
    exposure to new experiences 104–6
    illegal importers 211
    illegal puppy farming 210–13
    introducing to existing dogs 258–61
    play and bite inhibition 92–3, 119
    puppy classes 105–6
    puppy industry 210–13
    socialisation 104, 105–6, 224
purebreeds 201–8
puzzle games 274

Q
quality time 159, 169
quick fixes myth 99–100

R
Raz 10
reactive dogs 171–97
    feeling of guilt over 245
    F.O.D. (fear of dogs) 180–1, 187
    F.O.P. (fear of people) 181–2
    proximity sensitivity 174–80
    reactive humans 189–92
    resource guarding 186–8, 244, 245–7
    rewarding 185–6
    when on a lead 183–4
reassurance 26, 99
recall, training good 109
reflexes 34
reinforcement 49–51, 185
    and poor behaviour 39
    positive 131, 191
rescue dogs 213, 215–37, 284
    body handling 107
    choosing 199
    'click and collect' dogs 226–7
    fear of strangers 116–18, 155, 156, 158, 160
    in-person home checks 221
    making a match 219–21
    overseas rescue dogs 222–4, 226
    'rule of three' 106
    safe spaces 143, 146–7
    senior dogs 225–6
    settling in 106–7, 108
    Sophie from Romania 160–2, 163–9
    taking home 228
resilience, building 119
resource guarding 41, 42, 186–8, 244, 245–7, 259, 266, 268, 285
respondent conditioning 35, 37–8, 39, 45
retrievers 202
retro pugs 207–8
rewards
    food rewards 110–12, 130–1, 132
    importance of timing 129–30
    play as 112
    randomising 132
    rewarding in advance 135–7
    rewarding bad behaviour misconception 185–6
Ripley 33, 43, 65–6, 126, 195–6, 201–2, 256
    and Murphy 259–60

INDEX

rivalry, sibling 264–9
role-reversals, in play 81, 123–4
Romanian rescue dogs 223
Rottweilers 205
routines 289
    familiar 255
    and life changes 257–8
Royal Guide Dogs Association of Australia 208
The Royal Society 55, 57
RSPCA 212, 217–18
Ruby (Dachshund) 193, 196
Ruby (Vizsla) 263–8, 269
Rugaas, Turid, *On Talking Terms with Dogs: Calming Signals* 75

S

Sadie 263–8, 269
safe spaces and bases 143–9, 152–69
St Bernards 205
salivary reflex 34
Salukis 203
San Francisco SPCA 14–15
scent hounds 203
Schenkel, Rudolph 89
Schnauzers 209
    Miniature Schnauzers 205
Scottish Society for the Prevention of Cruelty to Animals 211
screening programmes 207
self-handicapping in play 81, 123, 124
senior dogs rescue dogs 225–6
senses, dogs' 126
sensitivity to sound 143–4
separation anxiety 93–4, 95, 154, 180
setbacks 134, 146
setters 202
shock collars 128–9
shutdown behaviour 150–1
sibling rivalry 264–9

sight hounds 203
signals
    calming signals 75
    intention signals 79–80, 84
    meta signals 78, 80, 123
    threat signals 74–5
Sigsworth, Jane 123
silver foxes 21–2
sit request 132
Skinner, B. F. 38
Skye Terriers 271–2
sleep, elderly dogs 274
smell, and human emotions 270
snarking 189
snarling 74, 85
socialisation
    early 100, 104, 105–6
    importance of early 45
    pedigree dogs 202
    older dogs 93, 274
    puppies 93, 224
    social isolation 26
Soft-coated Wheatens 204
Sophie from Romania 160–2, 163–9, 222
sound sensitivity 143–4
spaniels 202
Spanish rescue dogs 223
spite myth 95
spoiling dogs 94
sporting dogs 202
Staffies 204
status symbols 24
strangers
    fear of 116–18, 168, 181–2, 266
    growling at 85
    stranger aggression 55
stress, smell of 270
stretches, fearful dogs and 165
stroking dogs 26, 69–70
Sudden Environmental Contrast (SEC) 236–7
syringomyelia 207

T

tails
  tucked tails 69, 70
  wagging tails 65, 73
teeth cleaning 107
terriers 204
Texas A&M 274
Thomson (Tyson) 5–14, 15–16, 134, 145–6, 216, 285–6
threat signals 74–5
thresholds 150
  moving 149–56
thunderstorms 143
timing, rewards and 129–30
tongue flicks 69, 73, 81
toy breeds 205–6
toys
  and life changes 258, 259
  resource guarding 90, 259, 268
training
  accidental trainers 103–12
  trainability 55, 57
  training snacks 109–10
  training with treats as bribery myth 94
treats
  and classical conditioning 45, 46–7, 48–9
  food rewards 110–12, 130–1, 132
  and life changes 258, 259
  rewarding in advance 135–7
  training with treats as bribery myth 94
triggers, fear 46, 47–8
trust conundrum 143–6
tug, as the wrong kind of play 96, 97–8
turning away 75
Tyson *see* Thomson (Tyson)

U

University of Auckland 262
University of California 261
University of Cambridge 206–7
University of Lincoln 269–70
University of Liverpool 223
University of Milan 271
University of Naples 270
University of Sao Paolo 269–70
University of Washington 274
upset dogs 36–40
US National Institute of Health 274
utility group 204–5

V

vets 241–2
  choosing 252–3
  vet visits 107
Victoria, Queen 206, 218
Vizslas 202, 263–8, 269
voice, tone of 25, 269

W

walking dogs 26–7
  reactive dogs 181, 193–4, 196
  walking to heel 109
Weimaraners 202
Westies 204
wheelbarrowing 177–8
whining 73
Whippets 203
Wilberforce, William 217
Winston 192–7
Wolf Science Center 21
wolf-dogs 19–21, 23, 56
wolfhounds 203
wolves 19–21, 23, 82, 88, 89–90
Woody 15–16, 260
work, taking dogs to 17–18
working dogs 55, 205
wrestling matches 81, 92

Y

yawning 26, 75
Yorkshire Terriers 57, 205